T0214670

A Brief History of Intelligence

F. Richard Yu • Angela W. Yu

A Brief History of Intelligence

From the Big Bang to the Metaverse

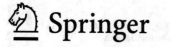

F. Richard Yu
Carleton University
Ottawa, ON, Canada

Angela W. Yu
Carleton University
Kanata, ON, Canada

ISBN 978-3-031-15953-4   ISBN 978-3-031-15951-0   (eBook)
https://doi.org/10.1007/978-3-031-15951-0

This Springer imprint is published by the registered company Springer Nature Switzerland AG
The registered company address is: Gewerbestrasse 11, 6330 Cham, Switzerland

Preface

A Brief Journey Through "A Brief History of Intelligence: From the Big Bang to the Metaverse"

How does intelligence arise?

Why intelligence has been evolving, from non-biological, plants, non-human animals to humans?

Can we make machines smarter than humans?

People have long wanted to answer the above questions, but so far they have not been satisfactorily answered. In recent years, the development of artificial intelligence (AI) has widely aroused people's interest in the phenomenon of intelligence and the nature of intelligence.

Dutch philosopher Baruch de Spinoza once said that the highest activity a human being can attain is learning for understanding because to understand is to be free. This book originated from my research and exploration to understand the phenomenon of intelligence and the nature of intelligence.

Although the progress of artificial intelligence in recent years has achieved exciting results in some fields, most of the current artificial intelligence research and development work is mainly concentrated in the field of engineering technology. The lack of understanding of the nature of intelligence limits the development of artificial intelligence. "You cannot solve this problem on the same level of thinking that caused it," Einstein said, "You have to go beyond it and reach a new level in order to solve this problem."

In the process of studying the phenomenon of intelligence and the nature of intelligence, our eyes cannot be limited to human intelligence. Instead, we should put our vision beyond human intelligence, consider different things in the universe, reach a new level, and study and explore the phenomenon of intelligence and the essence of intelligence on a new level.

This book introduces a variety of intelligence phenomena starting from the birth of the universe, including intelligence in physics, intelligence in chemistry, intelligence in biology, intelligence in humans, and intelligence in machines.

It uncovers the mystery of intelligence to the world and explores the natural phenomenon of intelligence. If understanding intelligence is regarded as a journey of a thousand miles, then this book is the first step to try.

By looking at the various phenomena of intelligence since the birth of the universe, we can see that intelligence is a natural phenomenon, similar to other natural phenomena (e.g., the rolling of rocks and the melting of snow and ice). These phenomena occur to facilitate the stability of the universe, and the phenomenon of intelligence is no exception.

This hypothesis can help us understand the nature of intelligence and explain all things in the universe, including plants, animals, humans, and the metaverse; they all have one thing in common: they play a role in the stabilization of the universe, and the phenomenon of intelligence happens naturally in the process. Different intelligent phenomena differ only in the dimensionality and efficiency in which they contribute to the stability of the universe.

The ideas in this book may offend the collective self-esteem of humanity and shake humanity's centrality to the universe. However, in the past history of human beings, the earth was expelled from the center of the universe by the Copernican revolution, and human beings were separated from the peak of biology due to the Darwinian revolution. So we shouldn't be particularly shocked when we learn that the human intelligence we're all proud of is actually similar to rocks.

The book is divided into 10 chapters, covering matter, energy, and space in the origin of the universe, gravity in physics, the principle of least action, dissipative structures in chemistry, entropy increase, maximum entropy production, the definition of life, the emergence of life, the intelligence in plants, the intelligence in animals, the neocortex structure of the brain, the special thinking of human beings, the theory of the brain, artificial intelligence symbolism, connectionism, behaviorism, artificial general intelligence, metaverse, etc.

Ottawa, ON, Canada   F. Richard Yu
Kanata, ON, Canada   Angela W. Yu

Contents

Chapter 1
Introduction

The highest activity a human being can attain is learning for understanding, because to understand is to be free.

— Baruch Spinoza

You cannot solve a problem on the same level that it was created. You have to rise above it to the next level.

— Albert Einstein

Humans have accomplished countless amazing things in our short history. We walked on the moon, created electronic devices, mastered flight, etc. With less than seventy thousand years, humankind has evolved from an insignificant animal in a corner of Africa to a species that stands on the verge of becoming a god, with the divine ability of creation. Many people have tried to use various theories and hypotheses explaining why we are the most intelligent species on the Earth. We have the sophisticated brain, nerve system, gossiping capability, language, etc.

But is it? The much less intelligent coronavirus can cause global pandemic, claiming more than 6 million intelligent human lives (as of May 2022). Ironically, the structure of most viruses is actually quite simple, nothing more than nucleic acid (DNA or RNA) wrapped around a protein subunit called a "capsid". The virus has no brain, no nerve system, even no blood, and even no complete cellular structure!

One may wonder who is more intelligent in this fighting, the coronavirus or humans? Of cause, we can design an effective vaccine to fight against the virus. However, the fact is that humans have already suffered great losses in the first round, and the virus is very likely to mutate and comeback again in the future.humans lose ground in the first round. Moreover, it is highly possible that the virus will comeback in the future.

Therefore, in this battle, who is more intelligent, the coronavirus or our humans? This is still a debatable question.

Looking back at the human history, we have seen that many large-scaled disasters were caused by viruses. The Black Death in the thirteenth century killed one-third of Europe's population, the Spanish flu in 1918 killed over 50 million people, and even today in the twenty first century, humans are still suffering from Ebola virus in

© The Author(s), under exclusive license to Springer Nature Switzerland AG 2023
F. R. Yu, A. W. Yu, *A Brief History of Intelligence*,
https://doi.org/10.1007/978-3-031-15951-0_1

2014 as well as the novel coronavirus which is still highly spreading now. Many lessons have been taught that the power of the virus cannot be underestimated. Viruses, the "lowest" life forms, have existed on earth for more than 4 billion years. By comparison, the short history of mankind, about 70,000 years, is but a drop in the ocean.

We may think, as a living thing, we are at least more intelligent than non-living thing. However, while human teleportation currently exists only in science fiction, teleportation is possible in tiny subatomic particles. Quantum teleportation allows two parties that are far apart to exchange information among them even in the absence of communication channel between them. Much more intelligent than humans!

One may argue that the capabilities of virus and quantum particles should not be called intelligence. Indeed, humans have been the primary focus of studying intelligence, e.g., in cognitive science and physiology of humankind. But recent studies show non-human animals and plants, even non-living things, exhibit intelligence.

What is intelligence? Although it is a concept that seems to have a concrete meaning in our daily lives, an abstract and quantifiable notion of intelligence is difficult to define. The word intelligence derives from the Latin nouns intelligentia or intellēctus, which in turn stems from the verb intelligere, to comprehend or perceive. The definition of intelligence is controversial, varying in what its abilities are and whether or not it is quantifiable.

The quest for intelligence has been around since as early as humankind coming to exist. Recently, the interest in intelligence has attracted heated attentions with the recent progress in artificial intelligence (AI).

Some people are excited to see that it is possible to create machines with human-like intelligence, helping us solve problems, such as autonomous driving, climate change and protein structure. For example, Google's Ray Kurtzweil envisions the Singularity, in which AI empowered by its ability to improve itself and learn on its own, will reach and then exceed human-level intelligence by 2040s [1].

On the other hand, some people are terrified by the progress and the advances of AI. For example, Elon Musk, founder of the Tesla and SpaceX companies, said that AI is probably "our biggest existential threat" and believes that "with AI we are summoning the demon."

Some prominent thinkers were pushing back, saying any reports of near-term superhuman AI are greatly exaggerated. Rodney Brooks, the former director of Massachusetts Institute of Technology (MIT)'s AI Lab, said that "we grossly overestimate the capabilities of machines—those of today and of the next few decades." Gary Marcus, a psychologist and AI researcher, stated that "general human-level AI has been almost no progress."

In studying human intelligence, it is usually related to the capacity for understanding, learning, reasoning, planning, creativity, critical thinking, and problem-solving.

Animal intelligence has also been studied in terms of problem solving, as well as numerical and verbal reasoning capabilities. In most cases, animal intelligence is often mistaken as biological instincts, or is determined entirely by genetics.

However, this is not always the case. Researchers have made a lot of observations and experiments to study animal intelligence.

For example, in an experiment, a banana was hanged on the top of the chimpanzee's cage, and a wooden box was put in the cage. After trying his best to catch the banana, the chimpanzee found the wooden box. After observation, he chose to put the wooden box under the banana, climbed up on the box, jumped vigorously from the box, and finally got the banana.

Plants are intelligent as well. We may naturally regard the plants as passively living things, but researchers have found that plants are not only capable of distinguish between positive and negative experiences and of learning from their past experiences, but also capable of communicating, accurately computing their circumstances, using sophisticated cost–benefit analysis and taking tightly controlled actions.

For instance, scientists have done research on semen cuscutae, a parasitic plant that does not perform photosynthesis. Scientists transplanted some semen cuscutae to some hawthorn trees with different nutritional status and found that semen cuscutae would choose to wrap around hawthorn trees with better nutritional status.

With hundreds of different definitions from psychology, philosophy, and artificial intelligence researchers, the definitions of artificial intelligence are as many as the experts trying to define it, said by Robert J. Sternberg. In general, the essence of intelligence in its general form can be referred as "an agent's ability to achieve goals in a wide range of environments" or "an agent's ability to actively reshape their existence in order to survive" [2].

In this sense, intelligence lies in not only living things, such as virus, but also non-living things, such as quantum particles. Nevertheless, it seems that, in common understanding, humans are more intelligent than non-humans, plants, and non-living things.

If you believe in Darwin's theory of evolution, you may naturally think that intelligence arises and develops through the natural selection. However, natural selection only explains *how* biological systems arise; it is difficult for natural selection to explain *what* characteristics they must possess, e.g., the active, such as the biological motivation, the end-directed purpose, the striving of living things (the "fecundity principle") and the increases in complexity in the absence of natural selection. Moreover, it cannot address the fact of planetary evolution, a special case of the problem of the population of one. Therefore, it is difficult to just simply use evolution to explain intelligence.

The Dutch philosopher Baruch de Spinoza once said, "The highest activity a human being can attain is learning for understanding, because to understand is to be free." This book originated from my research and exploration in order to understand intelligence.

I believe that intelligence is a natural phenomenon, las natural as rolling of rocks and the melting of snow and ice. Intelligence, like many other phenomena, can be studied by establishing simplified models. If intelligence is a natural phenomenon, can we answer the following questions?

- How did intelligence begin?
- Why has intelligence been evolving, from non-living things, plants, non-humans, to humans?
- Can we build machines that are more intelligent than humans?
- How to measure intelligence?
- Can we understand different forms of intelligence as completely, rigorously, simply as possible?

"You cannot solve a problem on the same level that it was created. You have to rise above it to the next level," said Einstein. We should not focus only on humans for studying intelligence. Instead, different things in the universe should be considered, and intelligence should be studied at a higher level.

If we study intelligence at a higher level, one possible hypothesis is that the natural phenomenon of intelligence, similar to other natural phenomena (e.g., rock rolling and ice melting), is *to make the universe more stable*.

I understand that the above point of view is rather dangerous, because it may offend the collective self-esteem of mankind, and knock us further off our pedestal of centrality in the universe. However, in our past history, Earth was dislodged from the center of the universe by the Copernican Revolution, and humans were yanked from the pinnacle of living things by the Darwinian Revolution. Therefore, it might not be so shocking for us to learn that human intelligence, which we all proud of, is actually similar to rock rolling.

Let me briefly explain this idea here. After the universe flashed into existence, the components in the universe are not unevenly distributed, resulting in a difference over a distance (e.g., in energy, mass, temperature, information, etc.). This difference is called *gradient*. Due to the gradient, the universe is not stable, and everything in the universe has never since been still. As stated by ecologist Eric Schneider, "nature abhors a gradient." Therefore, each component in the universe is contributing to relieving the imbalance through its own manner to the make the universe more stable. In addition, the stabilizing process of each component occurs in a distributed manner, not in a centralized one. Some simple examples include rock rolling and ice melting in our daily lives. Other sophisticated examples include living things evolution, collective intelligence, social networks, metaverse, etc.

This hypothesis can explain that all the things in the universe, including rocks, plants, animals, and humans, all have one thing in common: they contribute to the process of stabilizing the universe, and intelligence appears naturally in the process. Then, why do we have different things in the universe? In different environments, there are different constraints that limit the capability of stabilizing the universe.

Each thing (e.g., a particle, rock, person, company, or society) performs its most efficient way under the constraints to relieve the imbalance, and consequently stabilizing the universe. From this perspective, the main differences among the major categories of different things in the universe are as follows.

- Matter: Relieving the imbalance of energy to make the universe more stable.
- Non-human living things: Relieving the imbalance of energy, matter, and limited information to make the universe more stable.

- Humans: Relieving the imbalance of energy, matter, and more information to make the universe more stable.

The stabilizing process involves in a series of "phase transitions," rather than a single step. A phase transition is a holistic change in the overall arrangement of a system's structure, and consequently, its function. The timeline of different things' appearance in the universe indicates that new comers have more complex arrangements, and can contribute to the stability of the universe in more dimensions with more efficiency, compared to older things in the universe. We explain these points in the rest of this book.

Reference

1. R. Kurzweil, *The Singularity is Near* (Viking, New York, 2005)
2. S. Legg, M. Hutter, A collection of definitions of intelligence, in *Advances in Artificial General Intelligence: Concepts, Architectures and Algorithms*. Frontiers in Artificial Intelligence and Applications, vol 157 (IOS Press, Amsterdam, 2007), pp. 17–24

Chapter 2
Stabilizing the Universe

The universe was born restless and has never since been still.

— Henri Rousseau

Intelligence is the ability to adapt to change.

— Stephen Hawking

2.1 The Universe Made of Matter, Energy, and Space Out of Nothing

The universe we live in is a general term for the vast space and the various celestial bodies with diffused matter that exist in it. Although the origin of the universe is an extremely complex issue, people have been persisting in exploring when and how the universe came into being. Until the twentieth century, there were two influential cosmological models of the origin of the universe appeared: one is *the Steady State Theory*, and the other is *the Big Bang Theory*.

On one hand, according to the Steady State Theory, the past, the present and the future of the universe are basically in the same condition, which is equal and constant in terms of structure and has no beginning or end in terms of time.

On the other hand, the Big Bang Theory believes that the beginning of the universe and the time are originated from a huge explosion in the universe. It is the explosion that caused the major galaxies. Meanwhile, the major galaxies and the entire universe are always in a process of constant change and development.

In the Big Bang Theory, around 13.8 billion years ago, the entire universe, in all its mind-boggling vastness and complexity, ballooned into being out of the nothingness that preceded it. Of course the critical question is raised: did God create the Big Bang to occur? We have no desire to offend anyone of faith. So we leave this question out of the scope of this book.

In 1927, the Belgian cosmologist and astronomer Georges Lemaitre firstly proposed the Big Bang hypothesis [1]. In the late 1920s, Edmin Hubble discovered the phenomenon of redshift, indicating that the universe is expanding. In the mid-1960s, Arno Penzias and Robert Wilson discovered the cosmic microwave background radiation. Those two discoveries give strong support to the Big Bang

© The Author(s), under exclusive license to Springer Nature Switzerland AG 2023
F. R. Yu, A. W. Yu, *A Brief History of Intelligence*,
https://doi.org/10.1007/978-3-031-15951-0_2

theory [2], mainstreaming the Big Bang theory into the formal recognition of the origin of the universe.

In the Big Bang theory, the entire universe, about 13.8 billion years ago, with its incredible vastness and complexity, expanded from its previous nothingness. When the explosion happened, there was a point, where the volume was infinitely small, the density was infinitely high, the temperature was infinitely high, and the curvature of space-time was extremely infinite, called the singularity. At the beginning of the explosion, matter can only exist in the form of elementary particles such as electrons, photons and neutrinos. The continuous expansion after the explosion of the universe caused a rapid drop in temperature and density. As the temperature decreased, atoms, nuclei, and molecules were gradually formed and recombined into ordinary gases. The gas gradually condensed into nebulae, which further formed into various stars and galaxies, eventually forming the universe we see today.

Of course a critical question is raised: did God create the Big Bang ? We have no desire to offend anyone of faith. So we leave this question out of the scope of this book.

Despite the vastness and complexity of the universe, it turns out that to make one you need just three ingredients: matter, energy, and space [3]. Matter is the stuff that has mass. Matter is all around us, in our rooms, beneath our feet, and out in space. Water, rock, food, and air on Earth. Massive spirals of stars, stretching away for incredible distances.

The second ingredient we need to build a universe is energy. We encounter energy every day, although you have not thought about it. We use energy to cook food, charge phones, and drive cars. In a sunny day, we can feel the energy produced by the Sun 93 million miles away. Energy permeates the universe, driving the dynamic, endlessly changing processes.

The third one we need is space, lots of space. Wherever we look at our universe, we see space, stretching in all directions.

According to Einstein's theory of relativity, mass and energy are the same physical entity and can be changed into each other in his well-known equation, $E = mc^2$, where E is energy, m is mass, and c is the speed of light. This reduces the number of the ingredients in the "cosmic cookbook" from three to two.

Although only two ingredients, energy and space, are needed to make a universe, the big question is where these two ingredients come from. At the heart of the Big Bang theory, it explains that energy and space are positive and negative, respectively. In this way, the positive and the negative add up to zero, which means that energy and space can materialize out of nothing.

A simple analogy can be used to explain this crucial concept. Imagine we want to build a hill on a flat land, and we don't want to carry soil or rock from other places. To build this hill, we can dig a hole on this fat land, and use the soil from the hole to build it. In this example, we make not only the hill, but also the hole, which is a negative version of the hill. The hill was in the hole, and it perfectly balances out in this process. In other words, the hill and the hole can materialize out of a flat land.

This is the principle behind what happened for the energy and space at the beginning of the universe. When the Big Bang produced a massive amount of energy, it simultaneously produced the same amount of negative energy, which is the space. The positive and the negative add up to zero.

2.2 The Restless Universe

After the universe flashed into existence, it is not as static as it appears. Everything in the universe is endlessly changing to make it more stable. Figure 2.1 shows the timeline of the universe evolution after the Big Bang.

Scientists believe that, in the first moments after the Big Bang, the universe was extremely hot and dense, with a lot of energy. As the building blocks of matter, the quarks and electrons were formed. These fundamental particles roamed freely in a sea of energy. Quarks and electrons had only a fleeting existence as a plasma because the annihilation removed them as fast as they were created. As the universe cooled, the quarks condensed into protons and nucleons after about one ten-thousandth of a second after the Big Bang. Within a few minutes, these particles stuck together to form atomic nuclei, forming the first atoms, mostly hydrogen and helium. The 73% hydrogen and 25% helium abundances that exists throughput the universe today comes from this period during the first several minutes.

The 2% of nuclei more massive than helium present in the universe today were created hundreds of thousands of years later. Electrons stuck to the nuclei to make complete atoms. These atoms gathered in huge clouds of gas due to gravity, and

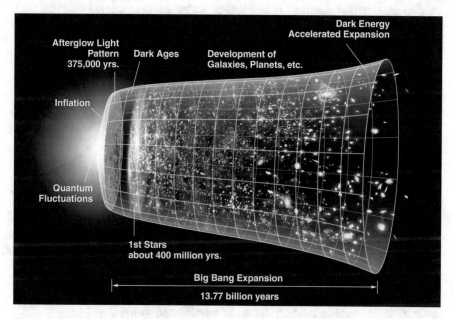

Fig. 2.1 The evolution of the universe after the Big Bang (Courtesy Wikipedia)

galaxies are formed with collections of stars due to gravity, which is the force that pulls any objects with mass towards one another, e.g., that causes falling apples from trees.

Large objects, such as Earth and the Sun, are in motion due to the gravitational force and the electromagnetic force. In addition to motion, the universe has been steadily expanding—increasing distances between galaxies embedded in space. An analogy to explain the expanding universe is a loaf of raisin bread dough. As the bread rises, the raisins on the bread move farther away from each other, but they are still stuck in the dough.

Between the year 1912 and 1922, American astronomer Vesto Slipher observed the spectra of 41 galaxies and found that 36 of them had a redshift, which he believed meant that these galaxies were away from Earth [4].

In 1929, observations by American astronomer Edmin Hubble showed that galaxies were moving away from Earth at a rate proportional to their distance, traditionally known as Hubble's Law. To commemorate Hubble's contribution, in 1990, the National Aeronautics and Space Administration (NASA) invented the space telescope, and named it "Hubble Space Telescope" to commemorate Hubble's great achievements.

In addition, the asteroid 2069 and the Hubble crater on the moon are all named after him. In 2018, the International Astronomical Union (IAU) voted to revise the name of the Hubble-Lemaitre law in recognition of the contributions of Hubble and Belgian astronomer Georges Lemaitre to the development of modern cosmology.

Small particles on the atomic and subatomic level, which are studied in the quantum world, are in motion as well due to the weak and strong nuclear forces. Not only do small particles move, but also they move strangely compared with that seen in our daily lives. Quantum particles can behave like particles, located in a single place; or they can act like waves, distributed all over space or in several places at once. Another strangest aspect of quantum particles is entanglement: if we observe a particle in one place, another particle—even far away—will instantly change its properties, as if these two are connected by a mysterious communication channel.

2.3 Change to Stabilize the Universe

Why is everything in the universe endlessly changing? Although it appears to be demonstrable and fundamental facts, this question is currently not fully answered by science.

One possible reason is that the two ingredients (i.e., energy and space) in the universe make it unstable from the beginning, everything in the universe is constantly changing to make the universe more stable since then. Moreover, due to the space ingredient, energy is vastly distributed in the universe, and it seems that there is no centralized control in this stabilizing process. Therefore, each component is contributing to the stabilizing process in a distributed manner.

This hypothesis can explain why matter in the universe was formed in the first place. Matter was formed to make the universe more stable by spreading energy to relieve the imbalance of energy. This matter-forming process was similar to the way steam condenses to liquid droplets as water vapor cools. The molecules in water vapor are more spread apart than those in water droplets. The change in density is accompanied by spreading energy. In a warm environment, water in the vapor state. The environment and water are in a stable state. When the environment's temperature drops, there is a gradient between the environment and water, and the system is not stable anymore. The environment is in a lower energy state than the water vapor. To make this system more stable, the density of water molecules changes to facilitate energy spreading. Consequently, the water changes from the gas state to liquid state. Similarly, in the matter-forming process, the structures of the particles changed to facilitate energy spreading.

Some other examples including rock rolling and ice melting in our daily lives. More sophisticated examples include living things evolution, collective intelligence, and trending ideas in social networks. We will elaborate this in the following chapters. Particularly, since we are mainly interested in intelligence in this book, we show that intelligence appears naturally in the process of stabilizing the universe, as natural as rock rolling and ice melting.

References

1. G. Lemaitre, Un univers homogène de masse constante et de rayon croissant rendant compte de la vitesse radiale des nébuleuses extra galactiques. Ann. Soc. Sci. Brux. **47** (1927)
2. E.L. Wright, Frequently asked questions in cosmology: what is the evidence for the big bang? *Ned Wright's Cosmology Tutorial. Los Angeles: Division of Astronomy and Astrophysics* (2013). https://www.astro.ucla.edu/~wright/cosmology_faq.html#BBevidence
3. S. Hawking, E. Redmayne, K.S. Thorne, L. Hawking, *Brief Answers to the Big Questions* (John Murray, London, 2020)
4. V.M. Slipher, Spectrographic observations of nebulae. Pop. Astron. **23**, 21–24 (1915)

Chapter 3
Intelligence in Physics

This most beautiful system of the Sun, planets and comets, could only proceed from the counsel and dominion of an intelligent and powerful Being.

— Isaac Newton

Nature's imagination far surpasses our own.

— Richard P. Feynman

After matter was formed in the universe, there rose a natural science, physics, which studies matter, its motion and behavior, and the related entities of energy and force.

Physics has become the research basis of other natural science disciplines. As a basic discipline of natural science, physics studies the most basic forms and laws of motion of all matter ranging from the universe to elementary particles. In general, physics focuses on the study of matter, energy, space, and time, especially their respective properties and the interrelationships between them.

In this chapter, the phenomenon of intelligence in physics and some wonderful phenomena that make the universe more stable at the physical level will be introduced. It can be clearly witnessed that intelligence emerges in the process of pushing the universe toward stability on a physical level.

3.1 The Perfect Physical World

The universe is far beyond beautiful and breathtaking. It is perfect—eerily, uncannily perfect. All sorts of physical constant—the speed of light, the charge of an electron, the ratios of the four fundamental forces (i.e., gravity, electromagnetism, weak and strong)—seem fine-tuned to create and run the universe.

As mentioned in Chap. 1 that the essence of intelligence in its general form can be referred to be an agent's ability to achieve goals. The physical universe does have its intelligence.

The neutron is 1.00137841870 times heavier than the proton, which is a bare hydrogen nucleus. This allows a neutron to decay into a proton, electron and

F. R. Yu, A. W. Yu, *A Brief History of Intelligence*, https://doi.org/10.1007/978-3-031-15951-0_3

neutrino—a process that determined the relative abundances of hydrogen and helium after the Big Bang and gave us a universe dominated by hydrogen.

If the neutron-to-proton mass ratio were even slightly different, we would be living in a very different universe. For example, with far too much helium stars would have burned out too quickly for life to evolve, or one in which protons decayed into neutrons rather than the other way around, leaving the universe without atoms. So, in fact, we wouldn't be living here at all—we wouldn't exist.

Of course, there might be other forms of intelligent life, not dreamed of even by writers of science fiction, that did not require the light of a star like the Sun or the heavier chemical elements that are made in stars and are flung back into space when the stars explode. Nevertheless, it seems clear that there are relatively few ranges of values for the numbers that would allow the development of any form of intelligent life. Most sets of values would give rise to universes that, although they might be very beautiful, would contain no one able to wonder at that beauty.

3.2 Gravity from an Intelligent Agent

One of the fundamental forces in the universe is gravity, by which all things with mass or energy, including stars, planets, galaxies, rocks, and even light, are brought toward one another. Gravity is responsible for many of the structures in the universe. The attraction of the original gaseous matter caused by gravity in the early stage of the universe made it to begin coalescing, forming stars, grouping together into galaxies.

In our daily lives, gravity gives weight to physical objects, and causes rocks to fall from mountains. Gravity holds the solar system together, keeping everything—from the biggest planets to the smallest particles of debris—in its orbit. The connection and interactions cause by gravity drive the seasons, ocean currents, weather, climate, radiation belts, and auroras.

Why is there gravity? Everybody experiences it, but pinning done why there is gravity in the first place is difficult. Although gravity has been successfully described with laws devised by Isaac Newton and later Albert Einstein, we still don't know how the fundamental properties of the universe combine to create the phenomenon. Newton stated in 1687, "Gravity must be caused by an (intelligent) agent acting constantly according to certain laws [1]." Before Newton, no one had heard of gravity, let alone the concept of a universal law.

Who is this intelligent agent? "Whether this agent be material or immaterial, I have left to the consideration of my readers," admitted Newton.

For more than 200 years, nobody truly challenged what the gravity intelligence might be. Probably, any possible challengers were intimidated by Newton's genius.

Einstein wasn't intimidated. In 1915, with no experimental precursors, Einstein had dreamed up an intelligent agent that causes gravity. According to his famous Theory of Relativity, gravity is a natural consequence of a mass's influence on space and time [2].

Fig. 3.1 The theory of relativity tells us that gravity is not a force, but a curvature of spacetime (Courtesy NASA)

Both Newton and Einstein agreed that space and time have dimension (e.g., space has width, length, and height, and time has length). But Newton didn't think that space and time can be affected by the objects in it. Einstein did. He theorized that gravity is just a natural outcome of a mass's existence in space and time. For space, gravity can warp it, bend it, push it, or pull it. For time, gravity can also warp it by speeding it up or slowing it down. Figure 3.1 shows that gravity is not a force, but a curvature of spacetime.

Using a trampoline game, we can visualize Einstein's gravity warp of space. Our mass causes a depression in the stretchy fabric of space in the trampoline. Roll a ball past the warp at your feet, and it'll curve toward your mass. The heavier you are, the more you bend space.

It is a commonplace to think of the Theory of Relativity as an abstract and highly arcane mathematical theory that has no consequences for everyday life. This is in fact far from the truth. This theory is crucial for the Global Positioning System (GPS), which is used for navigation to your destination. The GPS system consists of a network of more than 20 satellites in high orbits around Earth [3].

A GPS receiver determines its current position by comparing the time signals it receives from the currently visible GPS satellites (usually 6–12) and trilaterating on the known positions of each satellite. The precision achieved is remarkable: even a simple hand-held GPS receiver can determine your absolute position on the surface of Earth to within 5–10 meters in only a few seconds. More sophisticated techniques (e.g., Real-Time Kinematic (RTK)), deliver centimeter-level positions with a few minutes of measurement, which can be used for high-precision surveying, autonomous driving, and other applications. To achieve this level of precision, the clock ticks from the GPS satellites must be known to an accuracy of 20–30 nanoseconds.

The Theory of Relativity predicts that we should see their clocks ticking more slowly. If this effect were not properly taken into account, a navigational fix based on the GPS would be false after only 2 minutes, and errors in global positions would continue to accumulate at a rate of about 10 kilometers each day! The whole system would be utterly worthless for navigation in a very short time.

3.3   Gravity and Dark Energy

Gravity alone can make the universe unstable, because gravitation is always attractive in both Newton's and Einstein's theories. If you take some matter and distribute it perfectly equally throughput space, this system is unstable, like a rock precariously balanced atop a thin spire.

So long as the conditions remain perfect, matter will stay uniform and the rock will remain balanced. However, if you give that rock the tiniest nudge, you will leave equilibrium. The same thing is true for the universe with only gravity, as the tiniest perturbation will lead to runaway gravitational growth in local volume of space achieving the greater density. Once this growth begins, it will never stop. This initially overdense region will grow to an even greater density, and attract matter towards it even more effectively. In fact, one can demonstrate that any initial, static distribution of matter at rest will collapse under its own gravity, leading inevitably to a black hole.

Einstein's initial solution was to add in something else: a cosmological constant. In his equation, a universe dominated by a cosmological constant would see the distance between any two points increase over time. In other words, gravitation works to attract masses towards one another, but the cosmological constant works to push any two points apart.

This isn't a satisfying solution. If you move a mass a little too close to another, gravitation overcomes the cosmological constant, leading to runaway gravitational growth; If you move a mass a little too far away, the cosmological constant overcomes gravitation, accelerating the mass away interminably.

In 1922, Alexander Friedmann derived the equations that govern how a Universe that was filled evenly. In other parts of the world, this same solution was derived by Georges Lemaître, Howard Robertson, and Art Walker. One of the wildest things about the solution is that it explicitly shows that the spacetime fabric of a universe cannot remain static. Instead, it must either expand or contract to make it stable. In our case, the universe is expanding.

The scientific consensus has been that we don't need a cosmological constant. We treat it as just another generalized form of energy with its own properties: dark energy. Einstein missed it because he insisted on a static universe, and invented the cosmological constant to achieve that goal.

Recent studies show that gravity could be an emergent phenomenon, not a fundamental force [4]. Specifically, gravity abide by the second law of thermodynamics, under which the entropy of a system tends to increase over time. Using statistics to consider all possible movements of the mass and the energy changes involved, scientists found movements toward one another are thermodynamically more likely than others. Moreover, the dark energy leads to a thermal-volume law contribution to entropy. In other words, gravity and dark energy are to make the universe more stable.

3.4 Least Action Principle: A Proof for the Existence of God

In 1744, Pierre-Louis Moreau de Maupertuis discovered the Least Action Principle in a bid to prove the existence of God [5]. Through a peculiar mathematical and theological reworking of Newton's laws of mechanics, he had anticipated acclaim. However, his arguments were ridiculed by intellectuals across Europe at the beginning.

This principle has turned out to be one of the most influential ideas in physics. By the end of the nineteenth century, the whole science of mechanics rested on this principle. It is not strange that the Least Action Principle is sometimes regarded as the greatest generalization in the realm of physical science. The principle remains central in modern physics and mathematics, being applied in thermodynamics [6], fluid mechanics [7], the theory of relativity, quantum mechanics [8], particle physics, and string theory and is a focus of modern mathematical investigation in Morse theory.

God, as an intelligent being, will act always by the most economical means, and therefore the "action" in any motion in the universe should be minimum. The Least Action Principle stated just that—that in any motion the action consumed (measured by the product of the mass, the velocity and the distance) would be a minimum. Nature is thrifty in all its actions thanks to the perfection of God.

Before the principle of least action was proposed, many similar ideas appeared in metrology and optics. When the rope stretcher in ancient Egypt measured the distance between two points, the rope fixed to the two points was tightened, which could reduce the distance to a minimum. Ptolemy emphasized in Chap. 2 of Book 1 of his Geographia that the surveyor must make appropriate corrections for errors in straight lines. The ancient Greek mathematician Euclid stated in "Reflection Optics" (Catoptrica) that when light is irradiated on a mirror, the incident angle of the reflected path of the light is equal to the reflection angle. Later, Hero of Alexandria proved that the length of this path is the shortest [9].

A simple way to approach this principle is to consider, in our daily lives, we always strive as hard as possible to save time and energy. In order to achieve this goal, we design tools, including computers and artificial intelligence. We believe humans are the most intelligent species on Earth because we can design tools to save time and energy.

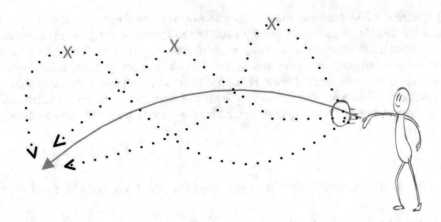

Fig. 3.2 The trajectory of a thrown rock follows the path of least action, in this case a parabola (red path). The dotted hypothetical course would require more action, which is defined as the total difference between potential and kinetic energy energies for the entire path. The Least Action Principle says that the path taken by the nature has the least action, which means the taken path is the most efficient one. In 1744, Pierre-Louis Moreau de Maupertuis discovered this Least Action Principle in a bid to prove the existence of God

However, Maupertuis found that, in the physical world, in all changes that take place in the universe, the sum of products of the speed of each body and the distance it moves is the least possible, as shown in Fig. 3.2. For example, if you throw a piece of rock, it'll find the most economical path back to Earth—that you can calculate its path by applying his principle. Maupertuis never doubted that he was on to something big. He titled his paper, "The laws of motion and rest deduced from the attributes of God." Then, in an odd inversion of that notion, he claimed to have constructed a proof for the existence of God.

The principle of least action can be intuitively understood from the name. The so-called action amount refers to a cost amount (Cost) that satisfies the above-mentioned principle to measure the choice of different motions. In classical mechanics, the amount of action refers to the amount of cost it takes to get from one point to another. Nature always chooses the path that minimizes this cost. In other fields, the specific form of this amount of action needs to be explored through experience. In different fields, the form of this action is different.

Light travels in straight lines, not curved trajectories, in a homogeneous medium. This is because the straight line distance is the shortest. However, light will be refracted in a non-uniform medium, because this can ensure that the reduced distance traveled by the light is the shortest, so that it takes the least time to arrive. That is, the way light is refracted is the route that makes the light travel the shortest time it takes to travel the optical path. This optical path is the amount of action in the light propagation process, defined as the product of the path length and the refractive index. This is Fermat's principle.

Another example is a slender chain, the ends of which are suspended at the same level. What is the shape of this chain? Intelligent nature will make the

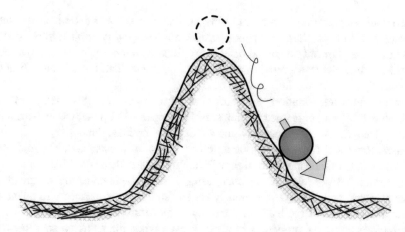

Fig. 3.3 A process of rock rolling down a valley. The system is not stable when the rock is on the high point of the valley. By taking the path of least action, the system stabilizes at a more efficient rate than would occur if taking another path. Intelligence (taking the path of least action) appears naturally in this stabilizing process

amount of action of this system—the gravitational potential energy—tend to be minimal. According to the variational principle, the shape of this chain, the so-called "suspension line", can be obtained.

Let's consider a simple example to explain this principle. Suppose a rock is placed on one side of a deep valley, as shown in Fig. 3.3. It will roll down the side of the valley. In this simple natural phenomenon, one explanation is to do with a balance between the forces of gravity and the forces of resistance provided by the air and surface. This is classical Newtonian physics. Another explanation, which dispenses with the notion of forces, relies on the fact that the system is not stable when the rock is on the high point of the valley, and there is a certain point on the contour of the valley in the stable state. In the stable state, their potential energy of the whole system (rock and Earth) is at a minimum, and that is on the very bottom of the valley.

In other words, rock rolling is a natural phenomenon that stabilizing the system. By taking the path of least action, the system stabilizes at a more efficient rate than would occur if taking another path. And intelligence appears naturally in this stabilizing process.

3.5 Quantum Teleportation: Spooky at a Distance

"Beam me up" is one of the most famous catchphrases from the "Star Trek" TV series. It's the command issued when a character wishes to teleport from a remote location back to the Starship Enterprise.

Human teleportation is currently only available in science fiction. In the subatomic world of quantum mechanics, teleportation is now possible, although it is not in the way typically depicted on VT. Specifically, teleportation in the quantum world involves the transportation of information, rather than the transportation of matter.

In quantum teleportation, the state of a particle can instantly "teleports" its state to two distant, entangled particles. In December 2020, scientists at Fermilab, a U.S. Department of Energy Office of Science national laboratory, and their partners presented for the first time a demonstration of a sustained, long-distance (44 kilometers of fiber) teleportation with fidelity greater than 90% [10].

Quantum teleportation is achieved using a natural phenomenon of quantum entanglement, which Einstein famously called "spooky at a distance". In quantum entanglement, one of the basic of concepts of quantum physics, the properties of one particle affect the properties of another, even when the particles are separated by a large distance. For example, if electron A and electron B are entangled, by changing something in one of the particles, it affects the other one instantly—in fact, even faster than the speed of light. You read it correctly,—faster than the speed of light. And no matter the distance between the two particles. Electron A can be on Earth and electron B on Jupiter. This is why such behavior had driven Einstein a bit mad. He thought it was spooky.

The fact that quantum particles can be entangled, makes a quantum computer more powerful than a classical computer. With the information stored in superposition, some problems can be solved exponentially faster. Developing a deeper understanding of entanglement can help solve both practical and fundamental problems. And entanglement may hold the key to some of the most fundamental questions in physics.

What is the cause of entanglement? There is no precise answer to date. Some researchers tried to explain it from the wave-function perspectives of the involved particles. Some researchers used the "Law of Conservation of Information" to explain quantum entanglement. In this line of research, it is believed that two particles are entangled because of being bound by a relationship/information, and information cannot be destroyed and conservation is satisfied at all instants. If something is changed in one of the entangled particles, the system is not stable, and another particle needs to change to make the system stable.

References

1. S. Chandrasekhar, *Newton's Principia for the Common Reader* (Clarendon Press, Oxford, 2003)
2. A. Einstein, Die feldgleichungen der gravitation, in *Sitzung der Physikalische Mathematischen Klasse.* **25**, 844–847 (1915)
3. GPS, Department o. global positioning system standard positioning service performance standard. GPS Augment. Syst. **35**(2), 197–216 (2008)

4. E. Verlinde, On the origin of gravity and the laws of newton. J. High Energ. Phys. **2011**(29), 1–24 (2011)

5. P. de Maupertuis, Accord différentes lois de la nature qui avaient jusqu'ici paru incompatibles. Leonhardi Euleri Opera Omnia 417–1744 (1911)

6. G. Vladimir et al., Thermodynamics based on the principle of least abbreviated action: Entropy production in a network of coupled oscillators, in Ann. Phys. **323**(8), 1844–1858 (2008)

7. C. Gray, Principle of least action. Scholarpedia **4**(12) (2009)

8. R.P. Feynman, The principle of least action in quantum mechanics. Ph.D. Dissertation, Princeton University, 1942

9. M. Kline, *Mathematical Thought from Ancient to Modern Times* (Oxford University Press, Oxford, 1972), pp. 167–168

10. L. Hesla, Fermilab and partners achieve sustained, high-fidelity quantum teleportation (2020) [Online]. Available: https://news.fnal.gov/2020/12/fermilab-and-partners-achieve-sustained-high-fidelity-quantum-teleportation/

Chapter 4
Intelligence in Chemistry

Chemistry stands at the pivot of science. On the one hand it deals with biology and provides explanations for the processes of life. On the other hand it mingles with physics and finds explanations for chemical phenomena in the fundamental processes and particles of the universe.

— Peter W. Atkins

Order arises from chaos.

— Ilya Prigogine

The story of intelligence continues with increasing levels of abstraction. With the appearance of rich information structure in atoms, such as carbon atoms, increasingly complex molecules were formed. As a result, physics gave rise to chemistry, and process of stabilizing the universe reaches a new level.

In the scope of its subject, chemistry occupies an intermediate position between physics and biology. Chemistry addresses topics such as how atoms and molecules interact via chemical bonds to form new chemical compounds, including their composition, structure, properties, behavior and the changes they undergo during a reaction with other substances.

The world is composed of matter, and there are mainly two forms of changes, chemical change and physical change (and nuclear reactions).

Unlike particle physics and nuclear physics, which study smaller scales, chemistry studies the interaction of atoms, molecules, and ions (clusters) of matter structure, chemical bonds, and intermolecular forces.

The scale of chemistry is the closest to the macro in the micro world, so their natural laws are also most closely related to the physical and chemical properties of substances and materials in the macro world in which humans live.

As an important bridge between the microscopic and macroscopic material world, chemistry is one of the main methods and means for human beings to understand and transform the material world.

Human life can be continuously improved and improved, and the contribution of chemistry has played an important role in it. We rely on chemistry to bake bread, grow vegetables and produce materials for everyday life. Chemistry is the basis for

© The Author(s), under exclusive license to Springer Nature Switzerland AG 2023
F. R. Yu, A. W. Yu, *A Brief History of Intelligence*,
https://doi.org/10.1007/978-3-031-15951-0_4

the formation of snowflakes, the science of champagne, the color of flowers, and other wonders of nature and technology.

In this chapter, we briefly review the development of chemistry, and then introduce some wonderful phenomena that make the universe more stable at the chemical level. We can see that intelligence emerges in the process of pushing the universe toward stability at the chemical level.

4.1   History of Chemistry Development

From primitive societies that started using fire to modern societies that use various man-made substances, we all enjoy the fruits of chemistry. Our ancestors drilled wood to make fire, used fire to bake food, drive away beasts, keep warm in cold nights, and make full use of the glowing and heating phenomenon when burning. It can be said to be the earliest chemical practice activity.

Combustion is a chemical phenomenon. After mastering fire, human beings have discovered some material changes one after another. For example, burning fire on copper ore such as emerald green malachite will generate red copper. The accumulation of these experiences and the formation of chemical knowledge led to social changes, the development of productive forces, the advancement of history, and the development of chemistry.

In the process of gradually understanding and utilizing the changes of these substances, human beings have manufactured products with great use value to human beings. Humans gradually learned to smelt and make pottery; later they learned to dye, brew and so on. These products processed and transformed from natural substances have become the symbols of ancient civilizations. On the basis of these exhaustive and productive practices, ancient chemical knowledge emerged.

From about 1500 BC to 1650 AD, chemistry accompanied the development of alchemy and alchemy [1]. In order to obtain the gold symbolizing wealth or the elixir of immortality, alchemists and alchemists did a lot of chemical experiments, and then books that recorded and summarized alchemy and alchemy appeared one after another. Although alchemists and alchemists all ended in failure, they explored the artificial transformation of a large number of substances in the process of "turning a stone into gold" and refining the elixir. At the same time, it has accumulated many phenomena and conditions of chemical changes in substances, and accumulated rich practical experience for the development of chemistry. The word "chemistry" appeared at that time, and its meaning was "alchemy".

Beginning in the sixteenth century, with the vigorous rise of industrial production in Europe, it promoted the creation and development of metallurgical chemistry and medicinal chemistry and turned alchemy and alchemy to life and practical applications. People began to pay more attention to the study of chemical changes in substances themselves. After the scientific concept of elements was established, the scientific oxidation theory and the law of mass conservation were established through precise experimental research on combustion phenomena. Subsequently, the law of definite ratio, the law of multiplication and the law of compound quantity were established, which laid the foundation for the further scientific development of chemistry.

The periodic table of chemical elements proposed by the Russian scientist Dmitri Ivanovich Mendeleev in 1869 greatly contributed to the development of chemistry [2]. Mendeleev arranged the 63 elements known at that time in the form of a table according to their atomic weights and put elements with similar chemical properties in the same row, which is the prototype of the periodic table.

Using the periodic table, Mendeleev succeeded in predicting the properties of elements not yet discovered at the time (gallium, scandium, germanium). In 1913, British scientist Moseler used cathode rays to hit metals to generate X-rays and found that the higher the atomic number, the higher the frequency of X-rays. Therefore, Moseler believed that the positive charge of the nucleus determines the chemical properties of the element.

He arranged the elements according to the positive charge in the nucleus (that is, the number of protons or atomic numbers), and after years of revision, it became the contemporary periodic table. The Periodic Table of Elements is what became the heart of chemistry.

Since the twentieth century, chemistry has developed from qualitative to quantitative, from macro to micro, from stable to metastable, and from experience to theory, which is then used to guide design and pioneering and innovative research. On the one hand, it provides as many new materials and new substances as possible for the production and technology sectors; on the other hand, new disciplines are constantly generated in the process of interpenetrating with other natural sciences, and develop towards the direction of exploring the origin of the universe and life sciences.

4.2   Dissipative Structures: Order Arises from Chaos

Look at the beautiful patterns shown in Figs. 4.1 and 4.2. You may wonder which intelligent artists designed these beautiful patterns? No, they were not deigned by humans. Rather, they were from the chemical interaction of some non-living substances. So, in terms of designing patterns, non-living chemical substances can be more intelligent than humans!

In 1900, Henri Bénard studied the first, and one of the most notorious examples of this pattern-forming phenomenon, in which the extraordinary regular array of convection cells that spontaneously develop, when a thin horizontal layer of fluid is heated uniformly from below as soon as the heat flux through the layer exceeds a well-defined threshold. Hexagonal patterns are the most familiar illustration, but simple rolls or squares are also possible. In each such cell, the hot fluid rises and when cooled at the top sinks back again to reheat.

In 1952, the British mathematician and computing pioneer Alan Turing realized that, if you mix some chemically reacting species, when some parameter crosses

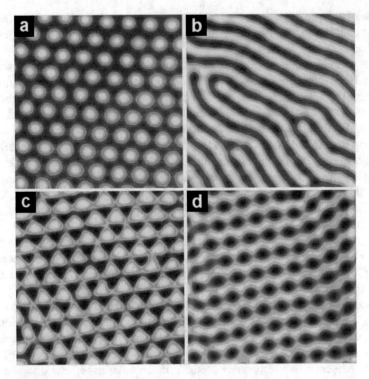

Fig. 4.1 (**a**)–(**d**) Turing structures of different symmetries obtained with the chlorite-iodide-malonic acid reaction. Dark and light regions, respectively, correspond to high and low iodide concentration. The wavelength, a function of kinetic parameters and diffusion coefficients, is of the order of 0.2 mm. All patterns are at the same scale: view size 1.7 mm × 1.7 mm (Courtesy P. De Kepper, CRPP)

a threshold value (e.g., the concentration of some chemical species), it leads to stationary, space-periodic patterns for the concentration of reactant, as shown in Fig. 4.1.

In 1950s and 1960s, two Russian scientists, Boris Belousov and Anatol Zhabotinsky, discovered the most famous of all oscillatory chemical reactions (it is now known as the Belousov-Zhabotinsky reaction or simply the BZ reaction). They found that in a mix of potassium bromate, cerium(IV) sulfate, malonic acid, and citric acid in dilute sulfuric acid, the ratio of concentration of the cerium(IV) and cerium(III) ions oscillate, causing the color of the solution to oscillate between a yellow solution and a colorless solution. In particular, the periodic variation of the concentration of the reaction intermediators and catalysts corresponds to a progressive variation of their geometry, form and color [3, 4]. Figure 4.2 just shows a single-shot picture of this dynamic process. If you are interested, you can search "Belousov-Zhabotinsky reaction" in Youtube, and you will find this beautiful phenomenon. The discovery sparked an intense debate in the field of applied physical chemistry.

Fig. 4.2 The Belousov-Zhabotinsky (BZ) reaction resulting in complex patterns in time and space. In a mix of potassium bromate, cerium(IV) sulfate, malonic acid, and citric acid in dilute sulfuric acid, the periodic variation of the concentration of the reaction intermediators and catalysts causes a progressive variation of their geometry, form and color (Courtesy National Geographic)

The creation of space-time structures is extremely interesting because self-organization order is created from an initial state of uniformity and chaos, and self-organization is directly related to intelligence.

Self-organization occurs in many physical, chemical, biological, robotic and cognitive systems. Particularly, interesting systems in the world around us—life, thought, combustion, ecosystems, traffic, epidemics, the stock market, the planetary environment, weather, cities—are also have these characteristic features, which emerge spontaneously in presence of a flux of matter, energy, and information [5, 6].

Ilya Prigogine published a research report "Structure, Dissipation and Life" at the International Conference on Theoretical Physics and Biology in 1969, and formally proposed the dissipative structure theory [7]. Prigogine was a Belgian physical chemist and theoretical physicist. Prigogine was born in Moscow on January 25, 1917. In 1921 he moved to Germany with his family. He settled in Belgium in 1929 and became a Belgian citizen in 1949.

The dissipative structure theory is a major achievement of the Brussels School of non-equilibrium thermodynamics and non-equilibrium statistical physics for more than 20 years. In establishing the theory of dissipative structures, Prigogine and his colleagues investigated the nature of spontaneously ordered structures such as B-Z chemical waves, Bennett convection, and chemical oscillatory reactions, as well as other biological evolutionary cycles. They use the concept of "self-organization" to describe the processes that form ordered structures, thus building a scientific bridge between "existence" and "evolution". Prigogine was awarded the 1977 Nobel Prize in Chemistry for this important contribution.

Prigogine believed that in a state of non-equilibrium without order, fluctuations in energy and matter can generate order from chaos [8]. The generation of spatial configurations and temporal rhythms in dissipative structures is a phenomenon known as "fluctuation order". He believes that modern science represented by Newton's classical physics describes a natural world like a clock, a static world that never develops, a world in which existence is absolute and relatively static. In classical Newtonian physics, changing the time parameter t to -t has the same result, time is reversible, and the past and future appear to be indistinguishable. However, according to recent thermodynamic results, as pointed out by the second law of thermodynamics, a closed system will only spontaneously increase its entropy and move towards random disorder (which will be introduced below). What this reveals is a world in which time has a direction and is constantly evolving. For example, the theory of biological evolution also tells us that the world is in constant development, and the arrow of time points to the future irreversibly.

The physical connotation of dissipative structure theory can be understood as: A non-linear open system (such as physical, chemical, biological and even social or economic systems) far from equilibrium states by constantly exchanging matter and energy with the outside world, the change of a certain parameter within the system reaches a certain threshold. At the same time, through fluctuations, the system may undergo sudden changes, that is, non-equilibrium phase transitions. From the original chaotic disordered state to a state of order in time, space or function. This new stable macroscopic ordered structure formed in the nonlinear region far from equilibrium is called "dissipative structure" because it needs to continuously exchange material or energy with the outside world to maintain it.

What is the root cause of dissipative structures? One possible reason is that, similar to the least-action paths discussed in Chap. 3, the dissipative structure enables the system to stabilize at a more efficient rate than if another structure (or no structure, i.e., chaos) were employed. Again, intelligence emerges naturally from this stabilization process.

Indeed, it is shown that the entropy of the system increases at a more rapid rate than would occur if the dissipative structure did not exist [9]. Generally speaking without technical detail, entropy increasing in the system means the system is changing from unstable state to more stable state. We will introduce the entropy concept in the next part. In this sense, "dissipative structures" should be called "stability-facilitating structures."

4.3   Entropy Increase: The Arrow of Time

Since intelligence is obviously related to "order," as shown in the above discussions, it is interesting to measure (i.e., quantify) "order" or "disorder" in a system. Entropy is a abstract concept to do this. The greater the entropy, the smaller the "order."

Fig. 4.3 (**a**) A clean kitchen with everything in order (**b**) A messy kitchen with nothing in order after several days. Entropy is used to measure "order." The entropy of (**b**) increases compared to that of (**a**)

Entropy is actually not mysterious. Like length and weight, it is used to measure things. Entropy is used to measure disorder, which is how chaotic something is.

The concept of entropy was introduced in 1865 by a German physicist Rudolph Clausius [10], one of the leading founders of the field of thermodynamics. The initial scope of thermodynamics was mechanical heat engines, which was extended to the study of chemical compounds and chemical reactions. Entropy is used in a variety of fields, from classical thermodynamics, where it was first recognized, to far-ranging applications in chemistry and physics, in biological systems and their relation to life, in cosmology, economics, sociology, weather science, climate change, and information systems including the transmission of information in our cell phones and the Internet [11].

A simple way to approach the concept of entropy is to consider a kitchen as an example. Suppose you clean your kitchen, and put everything in order. After several days, if you don't clean it, your kitchen will be in a mess, because you casually leave things here and there, and eventually nothing is in order, as shown in Fig. 4.3. You don't need to blame yourself about this. It is the nature to be blamed.

The second law of thermodynamics says that the entropy of any isolated system always increases. Isolated systems spontaneously evolve towards equilibrium—the state of maximum entropy of the system. More simply put: the entropy of the universe (the ultimate isolated system) only increases (or at least stay the same) and never decreases. Stephen Hawking said "The increase of disorder or entropy is what distinguishes the past from the future, giving a direction to time" [12].

As you read this book, entropy is all around you. The heat from your coffer is spreading out, cells within your body are dying and degrading, the floor is getting dusty, a coworker is making a mistake, crimes are occurring, news are coming from different sources, etc.

Entropy is fundamentally a probabilistic concept. Since a system usually consists of many components (e.g., cells within your body, items in your room, and molecules in your coffee). For every possible "usefully ordered" state of the system, there are many, many more possible "disordered" states.

We can describe it with a simple mathematical calculation. Suppose there are 20 items in the kitchen in Fig. 4.3, and there are 50 places where items can be placed. With mathematical knowledge of the permutations, we can calculate the total likelihood of placement:

$$C_{50}^{20} = \frac{50!}{20! \times 30!} \approx 4.71 \times 10^{13}. \qquad (4.1)$$

If "order" is defined as the state that each item is placed at a specific location and "disorder" is defined as all other states, then the probability of "order" occurrence is very low, almost impossible. By contrast, "disorder" will occur for sure. Therefore, it is very easy for the "order" state changes to the "disorder" state.

$$\text{The probability of "order"} = \frac{1}{4.71 \times 10^{13}} \approx 0. \qquad (4.2)$$

The higher the entropy, the more likely it is to happen. And the entire universe is spontaneously developing in the direction of greater possibility, that is, the direction of greater entropy. Therefore, the law of increasing entropy can be restated in the following way: it is possible for a state to evolve into a more probable state, that is, a more stable state.

Stated this way, the second law becomes almost a trivial statement. Here, it is assumed that the relative probability of a state is determined by the number of ways you can construct it from its elementary components. For example, there is only one way to put the molecules of a gas into one location in the corner of the room, but there are many ways to distribute them evenly, so they are all spread out. That means that the clustered molecules are probably going to evolve into the evenly-spread molecules when the time passes, hence the entropy increases.

In recent years, to make the concept of entropy easy to understand in chemistry and physics, there has been a shift away from the words "order" and "disorder," towards such as "spread" and "dispersal." In these systems, what entropy measures is how much energy is spread out in a process or how widely spread out it becomes. From the probability perspective, there are more ways for energy to be spread out than for it to be concentrated. Thus, the energy is spread out. Eventually, the system arrives at a state of called "thermodynamic equilibrium" with the maximum entropy, in which energy is uniformly distributed, and the system is stable.

From the "gradient" perspective, there is a difference over a distance (e.g., in energy, temperature, mass, information, etc.) in non-equilibrium systems. Due to the gradient, the system is not stable. For example, there is a temperature difference between a cup of hot coffee and the surrounding environment. The cup of hot coffee will eventually have the same temperature with the room it sits in. In addition, as long as the system is left alone, this process is irreversible. The coffee will never heat up again.

4.4 Maximum Entropy Production

The tendency of the entropy to a maximum as an isolated system (the second law of thermodynamics) has been known since the mid-nineteenth century. In terms of entropy production, this means that the entropy production is greater than or equal to 0.

Recently, independent theoretical and applied studies shown the maximization of the entropy production [13]. This principle is called maximum entropy production principle (MEPP). MEPP obviously represents new additional statement meaning that the entropy production is not just positive, but tends to a maximum. Thus, apart from the direction of the evolution, which follows from the second law's formulation, we have information about the movement rate of a system.

Similar to the least action principle described in the last chapter, this MEPP shows another example of nature taking the easiest and most accessible paths, and hence, processes are accomplished very quickly in a minimum time. The universe develops such as to achieve the final state as quickly as possible, and the appearance of ordered system is more efficient for realization of this process. Again, intelligence appears naturally in this process.

MEPP is confirmed in studies of various systems of the physical, chemical or biological origin on different observation scales (both microscopic and macroscopic), including atmosphere, oceans, crystal growth, transfer of electrical charge, radiation, biological evolution. For example, principles similar to MEPP appeared in theoretical biology a long time ago too. In 1922 Alfred J. Lotka proposed that evolution proceeds in such direction as to make the total energy flux through the system a maximum compatible with the constraints [14]. In other words, species, which utilize portions of the flow of available energy most efficiently (all other things being equal) for their growth and existence will increase their population and, therefore, the flow of energy through the system will increase.

References

1. Alchemy Lab, History of alchemy. Nature (2011)
2. Western Oregon University, A brief history of the development of periodic table, in *Chemistry 412 Course Notes* (2015). https://www.celinaschools.org/Downloads/Brief_Hist_Per_Tbl.pdf
3. B.P. Belousov, Периодически действующая реакция и ее механизм. Сборник рефератов по радиационной медицине **147**(145), 145–147 (1959)
4. A.T. Winfree, The prehistory of the belousov-zhabotinsky oscillator. J. Chem. Educ. **61**(8), 661 (1984)
5. S. Camazine, *Self-Organization in Biological Systems* (Princeton University Press, Princeton, 2003)
6. Feltz, B. Self-Organization, Selection and Emergence in the Theories of Evolution. In: Feltz, B., Crommelinck, M., Goujon, P. (eds) *Self-organization and emergence in life sciences.*, **331**, Springer, Dordrecht. (2006). https://doi.org/10.1007/1-4020-3917-4_20
7. I. Prigogine, Structure, dissipation and life, in *Theoretical Physics and Biology* (North-Holland Publishing Company, Amsterdam, 1967), pp. 23–52

8. P. Ilya, Time, structure, and fluctuations. Science **201**(4358), 777–785 (1978)

9. Y. Demirel, V. Gerbaud, Chapter 1: fundamentals of equilibrium thermodynamics, in *Nonequilibrium Thermodynamics*, 4th edn., ed. by Y. Demirel, V. Gerbaud (Elsevier, Amsterdam, 2019), pp. 1–85

10. S.G. Brush, *The Kind of Motion We Call Heat: A History of the Kinetic Theory of Gases in the 19th Century* (Elsevier, Amsterdam, 1976)

11. A. Wehrl, General properties of entropy. Rev. Mod. Phys. **50**(2), 221 (1978)

12. S. Hawking, *A Brief History of Time* (Bantam Dell Publishing Group, New York, 1988)

13. L. Martyusheva, V. Seleznevb, Maximum entropy production principle in physics, chemistry and biology. Phys. Rep. **426**, 1–45 (2006)

14. A.J. Lotka, Contribution to the energetics of evolution. Proc. Natl. Acad. Sci. U. S. A. **8**(6), 147 (1922)

Chapter 5
Intelligence in Biology

Biology is the study of complicated things that give the appearance of having been designed for a purpose.

— Richard Dawkins

Intelligence is based on how efficient a species became at doing the things they need to survive.

— Charles Darwin

Earth remains the only place in the universe known to harbor life. The earliest time that life forms first appeared on Earth is at least 3.77 billion years ago, possibly as early as 4.41 billion years—not long after the oceans formed 4.5 billion years ago, and after the formation of the Earth 4.54 billion years ago. As a result, chemistry gave rise to biology.

Biology is a science that studies the structure, function, occurrence and development of living things (including microorganisms, plants and animals). There are an estimated 2–4.5 million species of living things on our planet. There are many more species that have become extinct, estimated to be at least 15 million as well. From the deep sea to the mountains, from the Arctic to the Antarctic, from the hot tropics to the cold tundra, there are creatures. Their lifestyles are varied and they also have a variety of morphological structures.

This chapter briefly reviews the process of exploring the basic question of "what is life", and then introduces some research on the question of "why does life exist". Next comes the phenomenon of intelligence in microorganisms, intelligence in plants, and intelligence in animals. We can see that in the process of promoting the stability of the universe at the biological level, biological intelligence came into being.

5.1 What is Life

In a world governed by the second law of thermodynamics, all isolated systems are expected to approach a state of maximum disorder, as discussed in the last chapter.

© The Author(s), under exclusive license to Springer Nature Switzerland AG 2023
F. R. Yu, A. W. Yu, *A Brief History of Intelligence*,
https://doi.org/10.1007/978-3-031-15951-0_5

However, life on earth maintains a highly ordered state, evolving from the most primitive acellular state to prokaryotes with cellular structures, and from prokaryotes to eukaryotic unicellular organisms. Then, according to different directions, the fungi kingdoms, plant kingdoms and animal kingdoms appeared. Plant kingdom from algae to naked ferns to ferns, gymnosperms, and finally angiosperms. The animal kingdom has evolved from primitive flagellates to multicellular animals, from primitive multicellular animals to chordates, and then to higher chordates–vertebrates. The fishes in the vertebrates evolved to amphibians and then to reptiles, from which mammals and birds were differentiated, and one branch of mammals further developed into higher intelligent creatures, which is man.

In a world governed by the second law of thermodynamics, all isolated systems are expected to approach a state of maximum disorder, as discussed in the last chapter. Since life approaches and maintains a highly ordered state, some argue that this seems to violate the aforementioned second law, implying that there is a paradox. (We can see that organisms develop and evolve continuously from unicellular to multicellular, from low to high, from simple to complex, and from aquatic to terrestrial. Some argue that this appears to violate the second law of thermodynamics, suggesting a paradox.)

It is not a paradox. Although entropy must increase over time in a closed system, an open system can keep its entropy low, by increasing the entropy of its surroundings. The biosphere is an open system. In 1944, a physicist Erwin Schrödinger argued in his monograph "What Is Life?" that this is what a living thing, from virus to human, must do.

The increase of order inside an organism is more than paid for by an increase in disorder outside this organism by the loss of heat into the environment. By this mechanism, the second law is obeyed, and life maintains a highly ordered state.

There is a concept of life "living on negative entropy" and extracting "order" from the environment to maintain the organization of the system. That is also the thermodynamic basis of life [1].

The increase in order inside the organism far outweighs the disorder outside the organism due to heat loss to the environment. Through this mechanism, following the second law of thermodynamics, life maintains a highly ordered state.

For example, a plant, absorbs sunlight, uses it to build sugars, and ejects infrared light, which is a much less concentrated form of energy. The overall entropy of the universe increases during this process. As the sun energy dissipates through plant, the system is more stable than that without the plant.

Once again, the highly ordered structure of the plant is maintained from decaying, and the intelligence of the plant appears naturally in this stabilizing process.

5.2 Why Does Life Exist

A profound and old question is "Was the emergence of life in the universe an improbable event, or an inevitable one?" In other words, did life occur a result of chance, or was it an predicable and inescapable consequence of natural phenomenon?

This issue has been debated for a long time, and there is no conclusion so far. But people have at least one confirmed fact that there are no special elements in the chemical substances that make up life. Whether flowers or ginseng, ants or elephants, or ordinary people or Einstein, the basic chemical elements that make up life are these four: carbon, hydrogen, oxygen, and nitrogen. A little bit of other elements are also needed, mainly phosphorus, sulfur, calcium and iron.

Some scientists believe that if the earth returns to its original origin and re-evolves the history of life on earth, the earth will produce completely new species; However, opponents believe that the evolution of life is largely the product of the development of earth conditions to a certain stage. Although there are differences, the difference is not too big.

5.2.1   Chemical Evolution Theory

In the mainstream scientific view, it is assumed that life on Earth was an improbable event, which resulted from an improbable molecular collision in a primordial soup with a bolt of lighting and a colossal stroke of luck. Life originated from a series of chemical evolution processes from inorganic to organic, from simple to complex under primitive earth conditions.

Biomolecules such as proteins and nucleic acids are the material basis of life. The origin of these living substances is crucial to the origin of life. This hypothesis holds that on the primitive earth without life, due to natural reasons, non-living matter produces organic matter and biomolecules due to chemical action.

Therefore, the question of the origin of life is first of all the question of the origin of primitive organisms and the early evolution of these organisms. In the process of chemical evolution, a class of chemical materials is first created, and then these chemical materials constitute general "structural units" such as amino acids and sugars. Living substances such as proteins and nucleic acids come from various combinations of these "building blocks".

This hypothesis rests on the idea that Darwinian evolution is the exclusive means of adaptation in nature, where the complexity and diversity can be explained by random genetic mutation and natural selection. Since adaptive change requires genes, the emergence of life must have been a result of chance, rather than an evolutionary process.

In 1922, the biochemist Alexander Ivanovich Oparin was the first to propose a hypothesis of chemical evolution [2]. He believed that some inorganic substances on the primitive earth, under the action of energy from solar radiation and lightning, became the first organic molecules. He proposed that in the original nutrient soup, macromolecules such as polypeptides, polynucleotides and proteins will condense into aggregates, and these aggregates immersed in salts and organic substances can continuously exchange energy and matter with the external environment. Through "natural selection", the catalytic equipment of metabolism has been perfected day by day, the cryptographic relationship between nucleotides and polypeptides has been gradually established, and finally the accumulation of quantity has led to a qualitative leap, and life has finally been born.

In 1953, American scholar Stanley Lloyd Miller (Stanley Lloyd Miller) conducted a simulation experiment to verify the hypothesis of Obalin for the first time with experiments [3]. Miller simulated the atmospheric composition of the original earth at that time and synthesized organic molecular amino acids through spark discharge and heating with hydrogen, methane, ammonia and water vapor. Following Miller's experiment, many other important biomolecules that make up life have been synthesized by simulating the atmospheric conditions of the primitive earth, such as saccharine, saccharine, deoxyribose, ribose, nucleoside, nucleotide, fatty acid, Yelin and Lipids, etc.

In 1965 and 1981, scientists in China artificially synthesized insulin and yeast alanine transfer ribonucleic acid for the first time in the world. The formation of proteins and nucleic acids is the turning point from inanimate to animate. Generally speaking, the chemical evolution process of life includes four stages: from inorganic small molecules to organic small molecules; from organic small molecules to organic macromolecules; from organic macromolecules to form multi-molecular systems that can self-sustain and develop; Molecular systems evolved into primitive life.

A difficult problem that cannot be well explained by the theory of chemical evolution is in the primitive earth environment before the origin of life, how did nature change from small biological molecules (amino acids, nucleotides) to biological macromolecules (proteins, nucleic acids). As "The Mystery of Life's Origin: Reassessing Current Theories" points out: "Our achievements in synthesizing amino acids have been obvious to all, but we have consistently failed in synthesizing proteins and DNA; the two formed strong contrast". Although, with the development of science today, we can synthesize the required biological macromolecules in the laboratory with great efficiency using machines, but the synthesis experiments in the pre-life environment are difficult to succeed [4].

5.2.2   The Inevitability of Life

The opposing view, called "inevitable life", assumed that there are factors that the random motion of atoms and molecules is constrained in such a way that inevitably guarantees the emergence of life when conditions permit. Biological systems can emerge because they more efficiently spread, or dissipate energy, thereby increasing the entropy of the universe. This process was similar to the "order arises from chaos" phenomenon in chemistry described in the last chapter.

In 1995, Nobel prize-winning biologist Christian René de Duve presented this view in his book "Vital Dust". After his book was published, scientists from the

Santa Fe Institute and MIT studying the origin of life are arguing that de Duve's position should be the reigning one.

In a majestic sweep and bold speculation, he presents an awe-inspiring panorama of life on Earth, from the first biomolecules to the emergence of the human mind and the future of our species. In his book, he rejects the idea that life arose from a series of accidents, nor does he invoke God, goal-directed causes, or vitalism, which sees living beings as substances inspired by the spirit of life.

Instead, in an extraordinary synthesis of biochemistry, paleontology, evolutionary biology, genetics, and ecology, he argues for a meaningful universe in which life and thought emerge inevitably and deterministically because of the conditions of the time [5]. Starting with a single-celled organism, similar to modern bacteria, that emerged 380 million years ago and all forms of life appear on Earth today, he charts seven successive epochs corresponding to increasing levels of complexity.

He predicted that our species might evolve into a "human hive" or planetary super-organism in which individuals would give up some freedom for the benefit of all; or, if Homo sapiens disappeared, he envisioned us being replaced by another Replaced by intelligent species. After the book was published, scientists at the Santa Fe Institute and MIT who study the origin of life argued that his position should be accepted.

In 2016, E. Smith and H. Morowitz theorized in their book that life on Earth first emerged due to inanimate matter being driven by energy currents produced by the planet's geothermal activity, similar to that occurring in volcanoes and inside the Earth's core [6]. Life was an inevitable consequence of free energy buildup, presumably in areas like the hydrothermal vents in the ocean.

By relieving the energy imbalance through more efficient dissipation, life was formed as a sort of channel, as natural as water flowing downhill. Just as a channel that is carved into the hillside by water flowing downhill becomes progressively deeper with time, a metabolic pathway that is carved by energy flows becomes reinforced and strengthened. Living things are just a more efficient way for nature to dissipate energy, relieving the energy imbalance, increase the universe's entropy, and consequently stabilizing the universe.

The self-organizing process that led to life involved in a series of "phase transitions", rather than a single step. A phase transition is a holistic change in the overall arrangement of a system's structure, and consequently, its function. We can think of the emergence of human cognitive revolution as a phase transition, in which Homo sapiens (our ancestors) set apart from other animals. With the series of phase transitions, living things have more complex arrangements, in particularly those arrangements that are better at relieving free energy and stabilizing the universe.

With the same "inevitable life" school of thought, a professor at MIT, Jeremy England, and his team outline a basic evolutionary process called "dissipative adaptation," inspired by Prigogine's foundational work. In their papers [7, 8], they showed exactly how a simple system of lifeless molecules, which are similar to those existed on Earth before life emerged, may reorganize into a unified structure that behaves like a living organism, when hit with continuous energy flows.

This is because the system has to dissipate all that energy to relieve the energy imbalance. A biological system, which metabolizes energy to function through chemical reactions, provides an efficient way to do this. The simulation results in their student visually depict how such a complex system can emerge from simple molecules, when energy flows through that matter. This is much like the whirlpool that inevitably emerges in a draining sink.

While simulations were used in England's study, experiments that actually used physical materials demonstrated the same phenomenon. In 2013, a group of scientists from Japan showed that simply shining a light (energy flow) on a group of silver nano-particles enabled them to assemble into a more ordered structure, which can efficiently dissipate more energy from the light [9].

In 2015, another experiment demonstrated a similar phenomenon in the macroscopic world [10]. When conduction beads were placed in oil and struck with voltage from an electrode, the beads formed intricate collective structures with "wormlike motion" that persisted as long as energy flowed through the system. The authors remarked that the system "exhibits properties that are analogous to those we observe in living organisms." In other words, under the right conditions, hitting a disordered system with energy will cause that system to self-organize and acquire the properties we associate with life.

This tendency could account for the internal order of not only living things but also many inanimate structures as well. Snowflakes, sand dunes and turbulent vortices all have in common that they are strikingly patterned structures that emerge in many-particle systems driven by some dissipative process.

5.2.3   Self-Replication

Self-replication (or self-reproduction) is another distinguishing feature of life, which drives the evolution of life on Earth. This feature can also be explained by the "inevitable life" hypothesis. A great way of dissipating more energy over time is to make more copies of yourself.

Scientists have already observed self-replication in non-living things. Philip Marcus and his team at the University of California, Berkeley, reported in Physical Review Letters [11], that vortices in turbulent fluids spontaneously replicate themselves by drawing energy from shear in the surrounding fluid. Michael Brenner, at Harvard, and his collaborators presented theoretical models and simulations of microstructures that self-replicate [12]. These clusters of specially coated microspheres dissipate energy by roping nearby spheres into forming identical clusters.

Based on the facts that both living things and non-living things can have internal order and can self-replicate themselves, we can see that the distinction between living and non-living matter is not sharp, all of which are just contributing to stabilizing the universe.

5.2.4   Fractal Geometry

In order to effectively stabilize the universe, intelligence will naturally arise. One of the most amazing structures in biological systems is fractal geometry. Many natural and biological systems in the objective nature have a self-similar "hierarchical" structure, and in some ideal cases, the hierarchy even has infinite levels. When we appropriately enlarge or reduce the geometric size of things, the structure of the whole hierarchy does not change. Many complex physical, chemical and biological phenomena are behind fractal geometry that reflects such hierarchical structures. Fractal geometry occurs in natural and biological systems where efficiency is required, such as capillary networks, alveolar structure, brain surface area, spikes, or the branching patterns of leaves on trees.

Fractal objects are complex structures built using simple programs that involve very little information. This has clear benefits for organisms, as they must achieve the most efficient structures in the most economical way to achieve multiple goals [13]. Surprisingly, it is possible to develop mathematical functions based on fractal geometry algorithms to simulate them.

A good example of fractal geometry is the Roman cauliflower, shown in Fig. 5.1, which is a complex work of art and a mathematical marvel. The entire head is made up of smaller heads that mimic the shape of the larger head, and each smaller head is made up of smaller, similar heads. It goes on, goes on, goes on... The cauliflower presents an arrangement of organs with many spirals nested in various scales.

In 1975, the mathematician Benoit B. Mandelbrot coined the term "fractal" [14]. The best way to describe a fractal is to think about its complexity; a fractal is a

Fig. 5.1 The fractal structure of the Romanesco cauliflower

shape that retains the same complexity no matter how much you "zoom in" or out of focus. Fractal geometry is a geometry that takes irregular geometric forms as its research object. Simply put, fractal is the study of infinitely complex geometry with self-similar structures.

In traditional geometry, we study integer dimensions, such as zero-dimensional points, one-dimensional lines, two-dimensional surfaces, three-dimensional solids, and even four-dimensional space-time. In contrast, fractal geometry studies non-negative real dimensions such as 0.83, 1.58, 2.72, log2/log3 (see Cantor set). Because its research objects are ubiquitous in nature, fractal geometry is also called "the geometry of nature". Fractal geometry is the inherent mathematical order beneath the complex surface of nature.

The generation of a mathematical fractal is based on an iterative equation, a recursion-based feedback system. There are several types of fractals, which can be defined in terms of exhibiting exact self-similarity, semi-self-similarity, and statistical self-similarity, respectively. Although fractals are a mathematical construct, they can also be found in nature, which makes them classified as works of art. Fractals have applications in medicine, soil mechanics, seismology, and technical analysis.

5.3 Intelligence in Microbes

5.3.1 Microorganism

Microorganisms are the general term for all tiny organisms that are difficult to see with the naked eye and can only be observed with the aid of an optical microscope or an electron microscope. Microorganisms include bacteria, viruses, fungi and a few algae. They have different shapes due to different environments. Among the creatures in the universe, microorganisms are the first to appear. With them, there are plants, animals, and even the current human beings. But even if there are humans, they are not extinct.

Microorganisms evolved from prokaryotes at the beginning to eukaryotes later. They never had a nucleus and evolved to have a nucleus. Microorganisms are simple in structure and reproduce extremely rapidly by division. Some microbes can even reproduce for dozens of generations in a day. They also metabolize quickly. It is this incredible rate of reproduction, combined with the low-demanding state of existence, that allows microorganisms to survive and be ubiquitous on our planet today.

Microorganisms are very important to humans, and one of their effects is to cause epidemics of infectious diseases. Among human diseases, many are caused by viruses. The history of microorganisms causing human disease is also the history of human struggles against them. Although great strides have been made in the prevention and treatment of diseases caused by microorganisms, new and recurring microbial infections continue to occur. So far, a large number of viral diseases have lacked effective therapeutic drugs. The pathogenic mechanism of some diseases is not clear. The abuse of a large number of broad-spectrum antibiotics has created a strong selection pressure, mutating many strains, resulting in the emergence of drug resistance and new threats to human health. Some segmented viruses can mutate through recombination or reassortment. The most typical example is the new coronavirus that began to circulate globally in early 2020. Viruses that people think are not "smart" have claimed the lives of more than 6 million "smart humans" (as of May 2022).

5.3.2 Intelligent Slime Mold

Microbes are very intelligent. For example, the slime mold (Slime mold), because as a single-celled organism, they display intelligence is unimaginable. John Tyler Bonner of Princeton University said of slime molds, "They are nothing more than a bag of amoeba wrapped in a thin mucus sheath, but they have the same characteristics as having muscles and nerves and ganglia. animals–that is, simple brains–have the same variety of behaviors." Not only can they navigate mazes, they can learn, and they can even simulate the layout of artificial transportation networks. And all of this is based on the premise that slime molds have no nervous system and no brain.

The intelligence of slime molds first got people's attention, starting with a famous slime mold labyrinth experiment. In 2000, Japanese scientists such as Nakagaki set up such an interesting experiment [15]. They cultivated the slime molds in a normal maze, and put some slime molds' favorite food, oatmeal, at the beginning and end of the maze. In the maze, there are a total of 4 routes of varying lengths that can be connected to the two oat food sources.

At the beginning of the experiment, the researchers found that the slime mold stretched its cytoplasm, covering nearly the entire plane of the maze. And in the complex labyrinth, they couldn't hinder their intelligence at all. As soon as the slime molds find food, they begin to slowly retract the excess, leaving only the shortest path left.

In the experiment, the slime molds seemed to have negotiated, and without hesitation, they chose the path that consumes the least amount of energy and can get food.

If you think slime molds aren't that great at navigating mazes, they're even more intelligent. The road conditions are countless times more complicated than walking through the maze, and it is not difficult for them to find the "optimal solution".

Based on the above experiment, researchers designed a new experiment in 2004 to test the slime mold. In the new experiment, the researchers randomly placed multiple food sources on a flat surface to test whether the slime mold could also find the optimal path to forage multiple food sources. In this question, the key is what kind of circuit should be established to ensure that the least amount of energy is consumed, and you can eat all these cereals? Ultimately, the slime mold lived up to its expectations. The network formed by them connecting the points is almost the optimal path in the project.

Don't think that finding the optimal path is easy, this problem can contain extremely complex combinatorial optimization problems. And the complexity of the problem increases exponentially with the number of nodes. So it's not hard to imagine how difficult it would be to design a transportation network in the real world. But the really powerful thing about slime molds is that they can take all aspects of the situation into account, and the path they find is not the shortest, but the best.

With the above two laboratories, the researchers further wondered whether the slime mold could design a more complex network, the railway network of the entire Tokyo area of Japan!

We know that the railway system in the Tokyo area is one of the most efficient and well laid out in the world. Engineering and technical personnel spend a lot of manpower and material resources to design. However, slime mold, a single-celled organism without a nervous system and no head, only needs dozens of hours to grow wildly, and it can repeat the efforts of engineers and technicians for decades.

In this lab, the researchers created a large flat container following the contours of the Tokyo area. In addition, according to the light-shielding properties of slime molds, lighting was used to simulate the surrounding terrain and coastline to limit the activity range of slime molds. Because the real rail network is hindered by terrain, hills, lakes and other obstacles.

The researchers then placed the largest piece of oatmeal in the center of the container, representing the location of Tokyo Station. The other 35 small pieces of oats are scattered in the container. These small oats correspond to 35 stations in the Tokyo rail system, as shown in Figs. 5.2 and 5.3.

At the beginning of the experiment, slime molds will try to fill the flat surface of the container, so as to explore new areas. After more than 10 hours of continuous exploration and optimization, the slime mold began to optimize the layout as if it had a little understanding. The pipes between the link oats will continue to strengthen, while some pipes that are not very useful for the link will gradually shrink and disappear. After about 26 hours of constant exploration and optimization, the slime molds formed a network that closely resembles the Tokyo area's rail network. The network formed by the slime mold is a replica of the Tokyo Railway, even more elastic than the real Tokyo Railway [16].

From the experimental laboratory approximation of highway networks in 14 geographic regions by Andrew Adamatzky of the University of the West of England and his colleagues around the world: Australia, Africa, Belgium, Brazil, Canada, China, Germany, Iberia, Italy, Malaysia, Mexico, the Netherlands, the United Kingdom and the United States [17].

Even more incredible is that the network formed by the slime mold is also highly self-healing. For example, as long as one of the food sources is removed, the entire network will be rearranged according to the previous "optimization" principle.

But so far, how the brainless slime mold completes this intelligent network is still an unsolved mystery. It is precisely because of the "brainless" but displayed

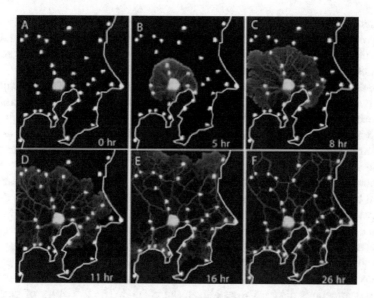

Fig. 5.2 Using slime molds to design the railway network in Tokyo, Japan

Fig. 5.3 The rail network formed by slime molds

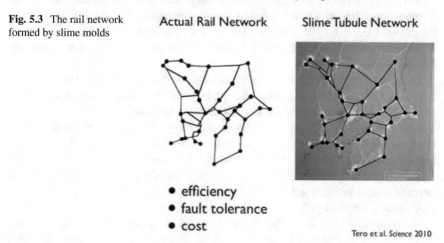

Actual Rail Network

Slime Tubule Network

- efficiency
- fault tolerance
- cost

Tero et al. *Science* 2010

wisdom that people wonder if this will be the key to opening the door to future artificial intelligence.

5.3.3 Tenacious Microorganisms

The researchers found that the same microbes "remind to communicate" with each other when their survival is threatened. John Woodland Hastings, one of the founders of chronobiology at Harvard University, proposed that if we could manipulate the transmission of information between these microbes, we could slow down the rate of microbial infection. This not only allows the patient to recover faster, but also does not allow the bacteria to produce antibodies.

"Bacteria are very intelligent organisms that can live anywhere and adapt very quickly to new environments," said the research team of Satish Nair, a Professor of biochemistry at the University of Illinois. Yersinia colitis, for example, as a food-borne virus can communicate through chemical signals, and they can respond together when the surrounding environment changes. Researchers are figuring out how to use these chemical signals to fight bacterial infections.

Another way to fight bacterial infections is to let one microbe kill another. British bacteriologist, biochemist, and microbiologist Alexander Fleming, discovered penicillin in 1928. Later, British pathologist Howard Florey (Howard Florey), German biochemist Ernst Boris Chain (Ernst Boris Chain) further research and improvement, and successfully used to treat human diseases, three People have won the Nobel Prize in Physiology or Medicine. The discovery of penicillin enabled humans to find a drug with a powerful killing effect on bacteria, ending the era when bacterial infectious diseases were almost untreatable. The discovery of penicillin also set off a climax of the search for new antibiotics, and since then mankind has entered a new era of synthesizing new drugs. Since then, various antibiotics have been developed, setting off a war of "killing" bacteria. Antibiotics disrupt the ability of bacteria to reproduce and grow by various means, preventing them from causing disease in humans.

In modern medicine, we use antibiotics all the time. Unfortunately, the smart thing about bacteria is that they can quickly adapt to antibiotics. So after an antibiotic is used many times, the bacteria become immune to it.

Nair has said that, in general, almost every type of bacteria is immune to at least one antibiotic. Researchers have discovered some "superbugs" that block all known antibiotics. This kind of bacteria can quickly develop drug resistance, because they "tell" other bacteria with chemical signals about the antibiotics they have been immune to, so that the same bacteria have developed drug resistance, which is why they become "superbugs".

Some researchers believe that the widespread use of antibiotics and the abuse of antibiotics are actually unscientific, because antibiotics are not good or bad, and they will kill good bacteria together. And surviving bacteria will develop antibodies to antibiotics and pass the antibodies on to other bacteria. In this way, the way to kill bacteria will only give rise to more powerful bacteria.

As the "lowest" life forms that are considered not "intelligent", why do microorganisms have such a large "intelligence" that they can defeat humans and exist on the earth for more than 4 billion years? From the perspective of promoting the stability of the universe, this is not difficult to understand. Everything a microbe does is maximizing its chances of survival and maximizing its offspring in the future. More offspring produce more entropy. If you look at the level of chaos in a child's room, you will understand the child's contribution to entropy production. As we saw in Chap. 4, the greater the entropy, the more likely it is to happen, that is, a more stable state. From this point of view, the intelligence of microorganisms is only a product that emerges as the times require in the process of promoting the stabilization of the universe.

5.4 Intelligence in Plants

In 1880, Darwin proposed the first modern concept of plant intelligence. In *The Power of Plant Movement*, he concluded that the roots of plants have "the power to direct the movement of adjacent parts." So "like the brain of a lower animal; the brain is located at the front of the body, receives impressions from the sense organs and directs several actions."

5.4.1 Advanced Sensory Systems

Plants have no eyes, but they can detect light. Although plants do not have noses, they can smell. Ripe fruit is a familiar technique in our daily life. Put a ripe apple or banana with a firm avocado or kiwi, and they will ripen very quickly. The reason behind this is that the unripe fruit smells the ethylene that the ripe fruit emits in the air.

In the 1930s, Richard Gein of the University of Cambridge demonstrated experimentally that, around ripe apples, the air contained ethylene. The Boyce Thompson Institute at Cornell University proposes that ethylene is a general-purpose plant hormone that ripens fruit. This mechanism ensures that the fruits of a plant ripen at the same time and are displayed together, like a fruit market for animals. What are the benefits of having the fruits of a plant ripen at the same time? The advantage is that after the animal finishes the fruit, it completes the task of helping the plant to spread the seeds so that the plant can reproduce and survive.

In addition to sight and smell, plants also have taste and touch. Animals use their tongues to taste food, and plant roots seek out the micronutrients they need in the soil, such as phosphates, nitrates and potassium. Carnivorous plants such as Venus flytrap and pitcher plant exist because of their demand for nitrogen. Carnivorous plants emit fragrant and sweet substances to trap their prey, and after they succeed, they produce enzymes to break down nutrients, which are absorbed by the leaves and metabolized by the captured animals. The sense of touch plays an important role in this.

Plants need to communicate with each other. For example, Liu Su said that many people like the smell of lawns. In fact, the volatiles that make up these smells are the alarm signals of grass. "It indicates that the leaf has been attacked by an external force (often insects in nature), so it is necessary to inform the neighboring grass blades to quickly synthesize some defensive chemicals. This is a way of

Fig. 5.4 Venus flytrap that "eats bugs"

communication between plants and can be considered a manifestation of plant intelligence."

Similar to the nervous systems of animals, plants can communicate with each other by sharing water and nutrients through an underground fungal network that sends chemical signals to other trees, alerting them to danger. In addition, plants can signal via gases and pheromones, electrical impulses in the air and underground. For example, when an animal starts chewing a plant's leaves, the plant can release ethylene gas into the soil, alerting other plants, and nearby plants can then send tannins into the leaves. So if they also chew the leaves, they might be able to poison the offending animal.

5.4.2 Intelligent Decision

Speaking of the intelligence of plants, many people should think of the Venus flytrap that can "eat bugs", as shown in Fig. 5.4. It is a perennial herb native to North America. Venus flytrap is a very interesting carnivorous plant. It has a "shell"-like insect trap on the top of its leaves and can secrete nectar. When a worm breaks in, it can be clamped at a very fast speed, and the worm will be eaten, digested and absorbed.

Every time the Venus flytrap is closed, it takes a lot of energy. If the prey caught is too small, the meat eaten is not consumed as much, even if it is caught, it will not be worth the loss. In order to achieve intelligent decision-making, the Venus flytrap can remember the stimuli it has received before, and even "count the seconds". Venus flytrap leaves have regular bristles on the edges, just like human eyelashes. A smart Venus flytrap doesn't sloppily close its clips for leaves falling from its side. If two of its trigger hairs are touched by an object within about 20 seconds, the blades will close, which means it will remember that one has been touched before and start counting the seconds. The Venus flytrap also remembers how many times the trigger hairs were triggered. After catching prey, the Venus flytrap will start secreting digestive juices after the trigger hair is touched 5 times.

It's not just quick-responding plants that make smart decisions. All plants respond to changes in their environment. They make decisions at the physiological and molecular levels all the time. In a water-deficient environment under the scorching sun, plants close their stomata almost immediately, preventing these tiny pores on the leaves from allowing water to escape. But is this response really "smart"?

When corn, tobacco and cotton are eaten by caterpillars, they produce chemicals that attract parasitic wasps. The parasitic wasp lays its eggs inside a caterpillar that eats the plant, and the caterpillar dies and feeds the wasp larvae.

The mallow orchid flower mimics the appearance and smell of female tonyd wasps in order to trick male wasps into pollinating themselves. Once the male wasp arrives, the mallard will "trap" it, and the wasp will then be covered in pollen and spread to another flower.

Plant intelligence goes beyond adaptation and response into the realm of active memory and decision-making. The 1973 bestseller The Secret Life of Plants, by Peter Tompkins and Christopher Bird, makes some wild claims. For example, plants can "read people's minds", "feel stress" and "choose" plant killers.

Monica Gagliano, Associate Professor of evolutionary ecology at the University of Western Australia, has done some interesting experiments with potted Mimosa pudicas. The mimosa is often called the "shame plant" or "don't touch me" because its leaves fold inward when disturbed. In theory, it defends against any attack, indiscriminately treats any touch or fall as an offense and shuts itself down. She published a study in 2014 saying shame plants "remember" that they're not actually dangerous falling from such a low height, and realize they don't need to protect themselves. She believes her experiments help demonstrate that "the brain and neurons are a complex solution, but not necessary for learning." She believes that plants are learning and remembering. By contrast, bees forget what they learn after a few days, while shame plants remember nearly a month [18].

If plants can "learn", "memory" and "communicate", then humans may misunderstand plants and ourselves. We must revisit our common understanding of intelligence.

5.5   Intelligence in Animals

Humans have long believed that we are the only intelligent species. Even if we admit that other animal species are intelligent, we are isolating humans from the entire animal kingdom. The famous Dutch psychologist, zoologist and ecologist primatologist Frans De Waal described the intelligence of various animals in the book "Wise and Spirit" [19].

Fig. 5.5   Tool-using chimpanzee

5.5.1   Using Tools

The ability to use tools is considered a uniquely human manifestation of intelligence. But some animals are also capable of making tools and using them. A species of chimpanzee in the Republic of Congo hunts with two branches of different lengths, the study found. One of the branches is a sturdy wooden stick about 1 meter long, and the other is a very flexible grass stem. As shown in Fig. 5.5, in the process of hunting ants, chimpanzees use long wooden sticks as shovels and dig a hole to lead to the ant nest. Then another flexible grass stalk is inserted into the ant hole, and the grass stalk is used as a bait. The ants bite the grass stalk, and then the chimpanzee pulls out the ant that bites the grass stalk like fishing and eats it. This combination of tools is extremely common. So the use of tools is not uniquely human intelligence.

In the process of using tools, some animals even rehearse the future situation of using tools in their minds, and then make effective action plans according to the rehearsed situation. In one experiment, zoologists put peanuts in a thin tube in a fixed position. For animals to get peanuts, they have to use something to push the peanuts out of the tube. In the experiments, the experimenters prepared capuchin monkeys with a variety of tools, ranging from long and short sticks to flexible rubber. After many wrong attempts, the monkeys finally chose the long stick and used the long stick to push the peanut out of the tube.

The zoologists increased the difficulty in later experiments, adding a hole in the middle of the tube. If the monkey pushed the peanut in the wrong direction with the tool, the peanut would fall into a jar, and the monkey could not get the peanut. After a series of failed attempts, the monkey discovered the pattern of this new experiment, pushed the peanut in the right direction with a long stick, and finally managed to get the peanut. This experiment is not easy. The same experiment is given to human children. Only human children after the age of 3 can successfully get peanuts.

Chimpanzees also participated in this experiment. The amazing thing is that they can get peanuts directly and successfully after thinking without trial and error like capuchin monkeys.

Not just mammals, but reptiles, birds, and even invertebrates have examples of tool use. New Caledonian crows can also combine tools. In an interesting experiment, the short stick was used to get the long stick, and then the long stick was used to get the food. Three of the seven crows successfully completed the task on the first attempt. In another experiment, clever alligators built a trap that used floating branches to attract waterbirds to rest on the branches, which then bobbed underwater. If there are few branches in the water, they go far away and pick branches to make traps. A type of coconut octopus near Indonesian waters, they cleverly carry their coconut shells home and use them as cover to move safely across the ocean floor.

5.5.2 Animal Language and Socialization

Some people think that the use of language is a unique talent of human beings. However, many animals can also use language to express their thoughts. The most common example is parrot learning, and some parrots are smart enough to use different vocabulary, which shows that parrots can connect ideas and language together.

In the ocean, dolphins are also intelligent creatures that can use language. Each member of the dolphin has its own distinctive language, which is a very high-frequency whistle sound. The young dolphin can make this whistle sound at the age of 1, and can indicate its specific identity since then. Sometimes these whistles are also imitated by other dolphins, and if the called dolphin hears it, it does respond. This case shows that animals also name each other and build their own social networks.

In the social network of animals, there will also be derivative cultural phenomena similar to those in the human social network. The researchers found that there are many interactive behaviors in the social network of chimpanzees, including cultural transmission behaviors, which ultimately make the whole group behave differently from other groups. They may even invent something called a "fashion," a popular action or game.

A group of chimpanzees in captivity, constantly changing their "fashion" behavior. During a certain period of time, the group of chimpanzees would line up in a column, trotting round and round a post at the same rhythm, dropping one foot gently and pressing the other hard. At the same time shaking his head, as if dancing. In another test, the experimenter played some intellectual games with the chimpanzees. If the same game was played repeatedly, the chimpanzees were distracted, and they became bored and tried to change the game with the experimenter.

References

1. S. Erwin, What is life? the physical aspect of the living cell. Am. Nat. **1**(785), 25–41 (1967)
2. A.I. Oparin et al., *The Origin of Life on the Earth*, 3rd edn. Academic Press, New York (1957)
3. S.L. Miller, A production of amino acids under possible primitive earth conditions. Science **117**(3046), 528–529 (1953)
4. C.B. Thaxton, W.L. Bradley, R.L. Olsen, The mystery of life's origin: reassessing current theories. Biochem. Soc. Trans. **13**(4), 797–798 (1984)
5. de Duve C, *Vital Dust: The Origin and Evolution of Life on Earth* (Basic Books, New York, 1995)
6. E. Smith, H.J. Morowitz, *The Origin and Nature of Life on Earth: The Emergence of the Fourth Geosphere* (Cambridge University Press, Cambridge, 2016)
7. T. Kachman, J.A. Owen, J.L. England, Self-organized resonance during search of a diverse chemical space. Phys. Rev. Lett. **119**, 038001 (2017)
8. J.M. Horowitz, J.L. England, Spontaneous fine-tuning to environment in many-species chemical reaction networks. Proc. Natl. Acad. Sci. **114**(29), 7565–7570 (2017)
9. S. Ito, H. Yamauchi, M. Tamura, S. Hidaka, H. Hattori, T. Hamada, K. Nishida, S. Tokonami, T. Itoh, H. Miyasaka, T. Iida, Optical assembly of highly uniform nano-particles by doughnut-shaped beams. Sci. Rep. **3** (2013)
10. D. Kondepudi, B. Kay, J. Dixon, End-directed evolution and the emergence of energy-seeking behavior in a complex system. Phys. Rev. E **91**, 050902 (2015)
11. P.S. Marcus, S. Pei, C.-H. Jiang, P. Hassanzadeh, Three-dimensional vortices generated by self-replication in stably stratified rotating shear flows. Phys. Rev. Lett. **111**, 084501 (2013)
12. Z. Zeravcic, M.P. Brenner, Self-replicating colloidal clusters. Proc. Natl. Acad. Sci. **111**(5), 1748–1753 (2014)
13. J. Calkins, *Fractal Geometry and its Correlation to the Efficiency of Biological Structures.* Technical Report, Honors Projects, https://scholarworks.gvsu.edu/honorsprojects/205, 2013
14. B.B. Mandelbrot, B.B. Mandelbrot, *The Fractal Geometry of Nature*, vol. 1 (WH freeman, New York, 1982)
15. T. Nakagaki, H. Yamada, Á.Tóth, Maze-solving by an amoeboid organism. Nature **407**(6803), 470–470 (2000)
16. A. Tero, S. Takagi, T. Saigusa, K. Ito, D.P. Bebber, M.D. Fricker, K. Yumiki, R. Kobayashi, T. Nakagaki, Rules for biologically inspired adaptive network design. Science **327**(5964), 439–442 (2010)
17. A. Adamatzky, S. Akl, R. Alonso-Sanz et al., Are motorways rational from slime mould's point of view? Int. J. Parallel Emergent Distrib. Syst. **28**(3), 230–248 (2013)
18. M. Gagliano, M. Renton, M. Depczynski, S. Mancuso, Experience teaches plants to learn faster and forget slower in environments where it matters. Oecologia **175**(1), 63–72 (2014)
19. de Waal F, *Are We Smart Enough to Know How Smart Animals Are?* (WW Norton & Company, New York, 2016)

Chapter 6
Intelligence in Humans

The brain is a three-pound mass you can hold in your hand that can conceive of a universe a hundred billion light years across.

— Marian Diamond

The brain is the last and grandest biological frontier, the most complex thing we have yet discovered in our universe.

— James Watson

It is believed that animals much like modern humans first appeared about 2.5 million years ago. About 70,000 years ago, the cognitive revolution occurred in a species called *Homo sapiens* in Africa. The brain structure of sapiens achieved a threshold of sophistication and capacity such that ideas, knowledge and culture were formed. Consequently, biology gave rise to history.

In this chapter, we introduce the efficient structure of the neocortex in the human brain. Next is the special way of thinking of human beings, the theory about the human brain. Finally, it discusses the deficiency of human intelligence in dealing with the problem of information overload and the phenomenon of information cocoon.

6.1   Neocortex in Brain: An Efficient Structure

What caused the cognitive revolution in sapiens? We do not know exactly. One thing we do know is that cognitive revolution enabled new ways of thinking and communicating among sapiens. Darwinians believe that random genetic mutations changed the inner structure of the brains of sapiens, making them more intelligent. But, why did it occur only in sapiens, not Neanderthals, who were driven to extinction by sapiens? We are not sure.

One possible reason is that the special structure of the brain is caused by the information currents, similar to the phenomenons of dissipative structures described in the previous chapters. There is no doubt that information is important for an

© The Author(s), under exclusive license to Springer Nature Switzerland AG 2023
F. R. Yu, A. W. Yu, *A Brief History of Intelligence*,
https://doi.org/10.1007/978-3-031-15951-0_6

animal to survive and thrive. At every moment, an animal is facing tremendous information, including food, water, shelter, predators, environments, etc. Driven by the information currents produced by the environments, a special structure, neocortex, occurred in mammals' brain.

Neocortex is from the Latin meaning "new rind". This structure enables the brain to relieve the imbalance between the information outside the brain and the information inside the brain at a more efficient rate than other structures. In other words, using this structure, the system (brain and environment) stabilizes at a more efficient rate than would occur if using another structure. And intelligence appears naturally in this stabilizing process.

The gray matter covered on the surface of the brain is called the cerebral cortex, which is the material basis of advanced neural activities and is composed of neurons, nerve fibers and glia [1]. The human cerebral cortex has a large number of folds called gyri, the shallow gaps between the gyrus are called grooves, and the deep and wide grooves are called fissures. The area of the sulcus gyrus increases the area of the cortex. The surface of the cerebral cortex is divided into five lobes–frontal, parietal, temporal, occipital, and limbic. The frontal, parietal, temporal, and occipital lobes appear later in the phylogeny and are called the neocortex, and the limbic lobes appear earlier and are called the old cortex.

The cerebral cortex is divided into six layers from outside to inside: the molecular layer, the outer granular layer, the pyramidal cell layer, the inner granular layer, the ganglionic cell layer, and the polymorphic cell layer, which are composed of different types of nerve cells. The granule cells receive sensory signals, and the pyramidal cells transmit motor information. According to evolution, the cerebral cortex is divided into the ancient cortex (archeocortex), the old cortex (paleocortex) and the neocortex. The ancient cortex and the old cortex are related to the sense of smell and are collectively referred to as the olfactory brain. In mammals, the higher the rank, the more developed the neocortex. The ancient and old cortex is a three-layer cortex, while the neocortex develops into six layers. Due to the highly developed human neocortex, it occupies about 96% of the entire cortex.

The neocortex is a hallmark of the mammalian brain and is not present in birds or reptiles. The neocortex is the largest part of the mammalian cerebral cortex. It is about 2–4 mm thick in the top layer of the cerebral hemisphere and is related to some higher functions such as perception, motor command generation, consciousness, spatial reasoning and language. It's called "neo" because it's evolutionarily the newest part of the cerebral cortex. It is also the most divergent part of mammalian species, as shown in Fig. 6.1. The size of the neocortex varies widely among different mammals. In rodents, it is about the size of a postage stamp and is smooth. In primates, the neocortex is intricately folded over the deep ridges, grooves and wrinkles of the rest of the brain to increase its surface area. Because of its careful folding, the neocortex forms the bulk of the human brain. About 80% of the weight of the human brain comes from the neocortex. The appearance of the large forehead of Homo sapiens gave Homo sapiens a larger neocortex.

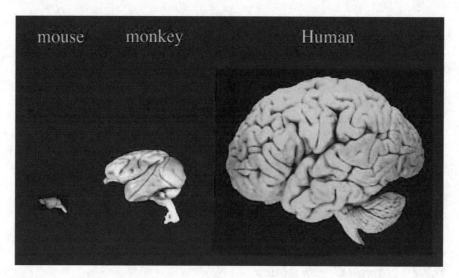

Fig. 6.1 Comparison of mouse, monkey and human brains

6.2 The Special Way of Thinking of Human Beings

6.2.1 Abstraction Levels and Patterns

The development of the neocortex does not just bring benefits. There is a tremendous cost due to the development of neocortex. Normal mammals with the weight of sixty kilograms have a brain size of 200 cubic centimetres on average. In contrast, modern sapiens have a brain size of 1200–1400 cubic centimetres. The first problem is that it is difficult to carry around this jumbo brain inside a massive skull. Another more important problem is fueling this jumbo brain. The brain accounts for only 2–3% of our total body weight. However, about 25% of our body's energy is needed to fuel the brain when the body is at rest. By comparison, only 8% of rest-time energy is needed for other apes' brains.

Due to the high cost, energy efficiency is very important. In order to save energy, neocortex uses "patterns" to deal with information and do so in a hierarchical fashion. Ray Kurtweild called this as the "pattern recognition theory of mind". It was found that animals without a neocortex (e.g., non-mammals) are largely incapable of understanding hierarchies. Understanding the hierarchical nature of reality is a uniquely mammalian trait due to the neocortex, which is responsible for sensory perception, recognition of everything from visual objects to abstract concepts, reasoning to rational thought, and language.

Processing logic needs much more energy than recognizing patterns in brains. Therefore, human being have only a weak capability to process logic, but a strong capability of recognizing patterns.

In 1978, a neuroscientist, Vernon Mountcastle, observed the extraordinary uniformity in the organization of the neocortex, hypothesizing that it was composed of a single mechanism that was repeated over and over again, and proposing the cortical column as that basic unit. This basic unit is a pattern recognizer, which constitutes the fundamental component of the neocortex. These recognizers are capable of wiring themselves to one another. This connectivity is not prespecified by the genetic code. Instead, it is created to reflect the patterns we actually learn over time.

In a human neocortex, there are about a half million cortical columns, each containing about 60,000 neurons. In total, there are about 30 billion neurons in a human neocortex. It is estimated that there are about 100 neurons in each pattern recognizer within a cortical column, and there are about 300 million pattern recognizers in a human neocortex.

6.2.2 Human Gossip Ability

There is a tremendous cost due to the development of neocortex. Normal mammals with the weight of sixty kilograms have a brain size of 200 cubic centimetres on average. In contrast, modern sapiens have a brain size of 1200–1400 cubic centimetres. The first problem is that it is difficult to carry around this jumbo brain inside a massive skull. Another more important problem is fueling this jumbo brain. The brain accounts for only 2–3% of our total body weight. However, about 25% of our body's energy is needed to fuel the brain when the body is at rest. By comparison, only 8% of rest-time energy is needed for other apes' brains.

Due to the high cost, energy efficiency is very important. In order to save energy, neocortex uses "patterns" to deal with information and do so in a hierarchical fashion. Ray Kurtweild called this as the "pattern recognition theory of mind" [2]. It was found that animals without a neocortex (e.g., non-mammals) are largely incapable of understanding hierarchies. Understanding the hierarchical nature of reality is a uniquely mammalian trait due to the neocortex, which is responsible for sensory perception, recognition of everything from visual objects to abstract concepts, reasoning to rational thought, and language.

Processing logic needs much more energy than recognizing patterns in brains. Therefore, human being have only a weak capability to process logic, but a strong capability of recognizing patterns.

In 1978, a neuroscientist, Vernon Mountcastle, observed the extraordinary uniformity in the organization of the neocortex, hypothesizing that it was composed of a single mechanism that was repeated over and over again, and proposing the cortical column as that basic unit [3]. This basic unit is a pattern recognizer, which constitutes the fundamental component of the neocortex. These recognizers are capable of wiring themselves to one another. This connectivity is not prespecified by the genetic code. Instead, it is created to reflect the patterns we actually learn over time.

Fig. 6.2 Sun bird–an ancient
Shu totem

In a human neocortex, there are about a half million cortical columns, each containing about 60,000 neurons. In total, there are about 30 billion neurons in a human neocortex. It is estimated that there are about 100 neurons in each pattern recognizer within a cortical column, and there are about 300 million pattern recognizers in a human neocortex.

Despite the cost brought by neocortex, this new structure enabled sapiens to not only develop verbal and written language, tools and other diverse creations, but also transmit information about things that they have never seen, touched, smelled or that do not exist at all.

Gossips, legends, gods, myths, and religions appeared for the first time on Earth, as shown by Yuval Harari in *A Brief History of Humankind*. While other animals could previously say, "Careful! A lion!", sapiens could say, "The lion is the guardian spirit of our tribe." [4]

Interestingly, almost every ancient human tribe had similar totem worship. The word totem comes from the Indian word "totem", which means "its mark", "its kin". In the eighteenth century, anthropologists discovered Indian totem worship in North America. The word totem is the language of a tribe of North American Indians, representing the sign or emblem of a clan. People living in that tribe believed that the totem was the ancestor and protector of the clan, so their totem was a special mark shared by the members of the clan. This is basically the same concept we now have about surnames. Totem markings just indicate blood ties between members of the same clan.

Sun worship and bird spirit worship are the two earliest worships in human society, and sun worship is almost integrated with bird spirit worship. Because in the primitive thinking of human beings, the sun is a flaming bird flying in the sky. Figure 6.2 is an ancient Shu totem, the sun bird.

What is quite surprising is that this sunbird was once a common worship object of all mankind. Luan or Luo in ancient China, Amaterasu in Japan, Laibird in ancient Egypt, Thunderbird in ancient America, Klaunos (Zeus) in ancient Greece, Gallolus in ancient India, and terinle bird, they are all sunbirds. And the names of the sunbirds are very similar in phonetics, such as "Luan" in China, "Lai" in ancient Egypt, "Lei" in ancient America, and "Gallo" in ancient India.

The ancestors living in our country also had other custom of totem worship in primitive society. In the "Historical Records", the exorcism tigers, bears, pixiu, etc. led by the Yellow Emperor are likely to be the remains of the clan totems preserved in the legend. Huangdi is also known as Youxiong, Shun's grandfather is called Jiao Niu, and the feudal lords are called You Jia, etc. There are various legends.

In addition, traces of totem worship have also been found in pre-Qin historical records and Confucian classics. For example, "Zuo Zhuan—The Seventeenth Year of Duke Zhao" contains "the Dahao clan is named after the dragon, so it is the master of the dragon and the dragon is named", which describes a clan with the dragon as the totem; The bird is suitable, so it is recorded in the bird, and the bird is named for the bird master." It is also a clan with birds as totems. "Book of History ● Gao Taomo" has the saying that "the phoenix comes to the ceremony" and "the beasts lead the dance", which means that many clans with birds and beasts as totems jointly support Shun as their leader.

"Book of Songs—Xuan Niao" "The mysterious bird was born to Shang."—The royal family of Shang used the Xuan Niao as a totem, indicating that they believed that the Xuan Niao was their ancestor.

In primitive beliefs, people of this clan are believed to be descended from a specific species. In most cases, considered to be related to an animal. Thus, totem belief has a relationship with ancestor worship. In many totem myths, it is believed that one's ancestors are derived from a certain animal or plant, or have a relationship with a certain animal or plant, so a certain animal or plant has become the oldest ancestor of the nation. For example, "The Mysterious Bird of Heaven's Destiny descended to the Shang Dynasty" ("Historical Records") Mysterious Bird became the totem of the Shang Dynasty.

6.2.3 Mitigating Information Imbalance to Facilitate Stability

Since Homo sapiens are social animals, social cooperation is the key to survival and reproduction. If Homo sapiens in a tribe discover a lion or a common enemy, there is an information imbalance among Homo sapiens in that tribe. It is crucial to alleviate this information imbalance by transmitting this information to other Homo sapiens as quickly as possible.

Before the Cognitive Revolution, Homo sapiens maintained kaleidoscopic relationships in a population of about a few dozen individuals. When a group becomes too large, its social order becomes unstable and the group splits. How can they agree to the rules, who should be the leader, who should eat first, or who should mate with whom?

In the wake of the cognitive revolution, Homo sapiens have an unprecedented ability to mitigate information imbalances efficiently, giving them the flexibility to collaborate in large numbers, millions, or billions. They have the ability to deliver messages about things that don't exist, such as tribal spirits, nations, LLCs, and human rights. This enables cooperation among large numbers of strangers and rapid innovation in social behavior.

Any large-scale human group, including nations, corporations, or churches, requires a common myth in the collective imagination of individuals. Some 2.5 billion people now believe in the biblical creation story, and billions of people are fighting climate change together. Likewise, the dollar exists in the shared imagination of billions of strangers.

This unprecedented cooperation benefits from the special structure of the human brain. And this special structure is produced under the drive of information flow, just as the flow of water produces a special valley structure, and the flow of energy produces a special life structure. These special structures allow the system to alleviate information, energy, and matter imbalances at a more effective rate than other structures.

In other words, humans use the special structure of the brain to stabilize the system more efficiently than other structures. Again, we see that intelligence emerges naturally from this stabilization process.

6.3 Theories About the Brain

For scientists studying intelligent machines, one obvious approach is to mimic human brains in computer programs so that human intelligence could be replicated in computing machines. They believe that a brain is a hunk of matter that obeys physical laws, and the computer can simulate anything. In order to do this, a theory about how the brain works is paramount important.

Despite the wealth of empirical data in brain and neuroscience, there are relatively few global theories about how the brain works. In this part, we present some of these theories, including the Bayesian brain hypothesis, the principle of efficient coding, and the free-energy principle.

6.3.1 The Bayesian Brain Hypothesistle

According to this hypothesis, the brain operates in situation of uncertainty in a fashion similar to Bayesian statistics [5]. Due to the ever-changing environments, humans and other animals' brains operate in a world of sensory uncertainty. The brain must effectively deal with the uncertainty to guide the correct actions. The underlying idea of this hypothesis is that the brain has a model of the world. When the sensory input signals come (e.g., when something is seen or something is heard), the brain actively explains and predicts its sensations. In this hypothesis, there is a probabilistic model that can generate predictions, against which the sensory input signals are compared. Based on the comparison results, the model is updated [6, 7].

In the eighteenth century, Thomas Bayes, an English theologian, mathematician, mathematical statistician and philosopher, and the founder of probability theory, came up with this neat, unremarkable Bayes theorem. This theorem was not published when he was alive, but it has played a huge role in various fields since then. Bayes' theorem is very simple, but this does not prevent it from becoming one of the hottest theories in contemporary cognitive science.

Bayes' theorem states that there are random events A and B, and the probability $P(A|B)$ of A occurring when B occurs is equal to the probability $P(B|A)$ of B occurring when A occurs, multiply by the probability of occurrence of A, $P(A)$, and divide by the probability of occurrence of B, $P(B)$.

$$P(A|B) = \frac{P(A)P(B|A)}{P(B)}. \tag{6.1}$$

Bayes' theorem allows us to infer the probability of something happening based on the known probabilities of related events happening.

When we got up in the morning, we looked at the weather, there were clouds in the sky, and we wanted to know how likely it was that it would rain today. Here, we can use Bayes' theorem to look at the probability of rain today.

Assuming known ahead of time:

- 50% of rainy mornings are cloudy!
- But there are actually quite a few cloudy mornings (about 40% of the days are cloudy in the morning)!
- This month is dominated by drought (on average only 3 days out of 30 days will rain, 10%)!

So what is the probability that it will rain today?

We use *rain* for rain today and *cloud* for cloudy morning.

When it is cloudy in the morning, the probability that it will rain that day is $P(rain|cloud)$.

$$P(rain|cloud) = P(rain) \times P(cloud|rain)/P(cloud), \tag{6.2}$$

where $P(rain)$ is the probability that it will rain today = 10%, $P(cloud|rain)$ is the probability of having a cloud in the morning on a rainy day = 50%, and $P(cloud)$ is the probability of cloudy morning = 40%.

The basic probability situation has been determined, it is simple.

$$P(rain|cloud) = 0.1 \times 0.5/0.4 = 0.125. \tag{6.3}$$

The chance of rain today is 12.5%.

In the 1860s, Hermann von Helmholtz showed in experimental psychology that the brain's ability to extract perceptual information from sensory data was modeled in terms of probabilistic estimation. The brain needs to organize sensory data according to the internal model of the outside would. Many mathematical techniques and procedures have been developed for the Bayesian brain hypothesis. For example, in 2004, David C. Knill and Alexandre Pouget used Bayesian probability theory to formulate perception as a process based on internal models. To use sensory information efficiently to make judgments and guide action in the world, the brain must represent and use information about uncertainty in its computations for perception and action. The brain is an inference machine that actively explains and predicts the outside world based on internal models.

The Bayesian brain hypothesis has been used in building intelligent machines, particularly machine learning algorithms, which will be elaborated in the next chapter.

6.3.2   The Principle of Efficient Coding

This principle suggests that the brain optimizes the mutual information between the perceptual information from sensory data and the internal model in the brain [8], under the constraints on the efficiency of these representations. Intuitively, mutual information measures the information that two random variables share. It measures how much knowing one of these variables reduces uncertainty about the other.

At its simplest, the principle of efficient coding says that the brain and neural system should encode sensory information in an efficient fashion. This principle has been applied in neuro-biology, contributing to an understanding for the nature of neuronal responses. It is effective in predicting the empirical characteristics of classical receptive fields and provides a principled explanation for sparse coding and the segregation of processing streams in visual hierarchies. It has been extended to cover dynamics and motion trajectories and even used to infer the metabolic constraints on neuronal processing [9–11].

6.3.3   Neural Darwinism

In Neural Darwinism, the emergence of neuronal assemblies is considered in the light of selective pressure. The beauty of neural Darwinism is that it nests distinct selective processes within each other. In other words, it eschews a single unit of selection and exploits the notion of meta-selection [12]. In this context, (neuronal) value confers evolutionary value (that is, adaptive fitness) by select-ing neuronal groups that meditate adaptive stimulus–stimulus associations and stimulus–response links. The capacity of value to do this is assured by natural selection, in the sense that neuronal value systems are themselves subject to selective pressure.

This theory, particularly value-dependent learning, has inspired "reinforcement learning", which is an important branch of machine learning algorithms. Reinforcement learning is concerned with how an intelligent agent takes actions in an environment in order to maximize its cumulative reward [13, 14]. Reinforcement learning is one of three basic machine learning paradigms, alongside supervised learning and unsupervised learning. Reinforcement learning is behind the famous AlphaGo, a computer program that can beat any human in the Go game. This will be elaborated in the next chapter.

6.3.4   The Free-Energy Principle

This principle was proposed by Karl Friston [15], a British neuro-scientist at University College London and an authority on brain imaging.

He has pioneered and developed one of the most powerful techniques to analyze the results of brain imaging studies and reveal patterns of cortical activity and relationships between different cortical regions. Friston proposed the free energy principle of the brain and wanted to perfectly explain the mechanism of how the brain works with thermodynamics.

What is the free energy law in physics? We cover that in Chap. 4. In other words, it is that any self-organized system in equilibrium tends to a state of minimal free energy.

What is free energy? Free energy refers to the part of the reduced internal energy of the system that can be converted into external work in a certain thermodynamic process. It measures the "useful energy" that the system can output externally in a specific thermodynamic process [16]. When a system with energy exchange with the outside world (a cup of hot tea on the table) is in equilibrium, the free energy is the smallest (water temperature drops, heat spreads), which refers to a state where the entropy is as large as possible. When the water temperature drops to room temperature, it is the most stable state. The minimum free energy is the law of the interaction between the system and the external environment under the second law of thermodynamics.

The learning process of the brain's cognitive system also conforms to the principle that free energy tends to be minimal.

Simply put, we can think of the brain as the cup of tea just mentioned. Its external environment and this glass of water have an energy interaction relationship, which corresponds to the brain's acquisition of external information (perception) through things like eyes and ears. This glass of water will become more and more room temperature, which corresponds to the brain exchanging information with the outside world like this glass of water. In this process, the information about the outside world in the brain is getting richer and richer. It is not only passively adopted, but also actively predicted and made behaviors.

In the principle of minimum free energy of the brain, the state of learning is to obtain a perceptual state that meets the expectations of the brain through continuous adjustment of behavior. And the internal state of the brain can more accurately match the changes in the external world, so that unexpected situations do not occur. Together these two parts minimize the free energy of the brain. The power of this principle is enormous, and it can tell you why you see a lot of what you want to see, even though you don't usually know it.

The free-energy principle tries to provide a unified framework to place the existing brain theories within this framework, in the hope of identifying common theses by unifying different perspectives on brain functions, including perception, learning and action [17].

This principle is motivated by the fact that the defining characteristic of biological systems is that they maintain their states and form in the face of a constantly changing environment. From the brain's point of view, the environment includes both the external and the internal milieu. It is essentially a mathematical formulation of how brains resist a natural tendency to disorder. In order to do this, the brain must minimize its free energy. In the formulation, free energy is an upper bound on surprise, which means that if the brain minimize free energy, it implicitly minimize surprise [18]. Here, surprise means an event with low probability. For example, "snowing on a hot summer day" would be a surprise.

A surprise will cause information imbalance between the environment and the internal model of the brain. For example, in the internal model of the brain, "snowing on a hot summer day" is highly impossible. If this event does occurs, there is an information imbalance, and the system is not stable.

Here we see that the cognitive model includes two aspects, one is the state of the external world acquired by perception and action, and the other is the update of the internal model of the cognitive process within the brain. This internal model keeps predicting the motivations behind each sense, and the underlying future changes, while the behavior itself tends toward those outcomes that are good for survival. The purpose of learning is to make the model of the internal state more accurate (prediction is accurate), and on the other hand, to obtain more evidence that is beneficial to survival for behavioral decision-making. If the model's predictions are not correct, behavioral decisions cannot lead to correct results [19].

In contrast, taking action to change the source of information consumes far less energy than updating an internal model. We mentioned above that our body needs about 25% of the energy to power the brain.

6.4 Information Overload and Human Cocooning

In the past, when the information source is scarce due to the lack of technologies (e.g., Internet), taking action to change the information source may not be an available option, and updating the internal model is the only option. Nowadays, due to the popularity of the Internet and cellphones, information is everywhere, it is much easier to change the information source than changing the internal model of the brain.

Today, thanks to the ubiquity of the Internet and mobile phones, information is everywhere. We have various apps on our mobile phones, and Facebook subscribes to dozens or even hundreds of official accounts. Information is pushed to us like a flood of beasts, overwhelming us. With the development of science and technology, the doubling cycle of information has been shortened, and the amount of information has increased geometrically. It is reported that in the past 30 years, the information produced by human beings has exceeded the sum of the information production in the past 5000 years.

In the era of information explosion, although information brings us a lot of knowledge, it also brings a huge impact. It may make us anxious, unable to concentrate on the things in the moment, or it may lead to information catastrophe caused by the excess of information [20, 21].

As we mentioned earlier, our ancestors evolved the neocortex to process more information than other animals. But this neocortex, which we are proud of, is obviously powerless in the age of information explosion.

So how can we make our brain system more stable, the feasible way is to change the source of information, because changing the source of information is much easier than changing the internal model of the brain.

This phenomenon has been well exploited by the "Feed"-based recommendation algorithm, which has penetrated almost all Internet products, e.g., Tiktok, browsers, photo Apps, etc. These products collect the browsing history, likes, tweets and comments. Then, they can derive your internal model of your brain. For example, an important Facebook post stated, "The goal of News Feed is to show people the stories that are most relevant to them." If you have the history of browsing vaccine conspiracy theories or you like the tweets related to vaccine conspiracy theories, the computer program will derive that, in your internal model of your brain, you believe in vaccine conspiracy theories. And more information about vaccine conspiracy theories will be recommended to you. By this way, since you don't have the information imbalance between the information source and your internal model, you don't have much surprise, and you will feel happy.

One of the results of using Feed-based recommendation algorithms is to form an "Information Cocoon", which is a concept put forward by Cass R. Sunstein, a professor at Harvard Law School who wrote *Infotopia: How Many Minds Produce* in 2006. This term indicates a phenomenon on the current Internet: when facing numerous information online, people tend to see only what they want to see, and the algorithm will select their preferred information to them, which ends up narrowing down their horizon just like a silkworm making a cocoon for itself [22].

As early as the nineteenth century, the concept of "information cocoon room" was proposed. The French thinker Alexis de Tocqueville has found that democratic societies are naturally prone to the formation of individualism, which will spread as equality of status expands.

According to Sunstein, the Internet constructs a "communications universe in which we hear only what we choose and only what comforts us." In the book, he referenced the work of MIT professor Nicholas Negroponte, who "prophecied the emergence of 'the Daily Me,' an entirely personalized newspaper in which each of us could select...perspectives that we liked." Simply put, it means that people will only follow what they are interested in, which will narrow their horizons in the long run. For some members of the general public in society, it is a real opportunity, but also a risk, sometimes with unfortunate consequences for business and society [23].

Sunstein vividly described the phenomenon of the "personal daily" in his writings. In the Internet era, with the development of network technology and the rapid increase of network information, we can choose the topics we pay attention to from the massive information at will. At the same time, newspapers and magazines can be customized according to their own preferences, and everyone has the possibility to customize a personal daily newspaper for themselves.

This "personal daily"-style information selection behavior can lead to the formation of network cocoons. When an individual is imprisoned in the information cocoon that he has constructed for a long time, over time, his personal life presents a kind of stereotype and program. For a long time, I have been in excessive self-selection, immersed in the satisfaction of my personal daily report, and lost the ability to understand different things and the opportunity to contact, and unconsciously created an information cocoon for myself.

The information cocoons is only an intermediate result. It will have a complex and far-reaching impact in many aspects related to information-politics, democracy, economy, entertainment, lifestyle, etc.

Living in an "information cocoon" makes it impossible for the public to think through, because their own preconceptions will gradually become ingrained.

Societal groups are divided. Such a narrow-mindedness will bring about various misunderstandings and prejudices. Precisely because news is freely available, the public must make choices in the face of countless news. If everyone only chooses the news they like to watch according to their own wishes, then everyone's world view is only what they want to see, not the way the world should have.

Living in an information cocoon for a long time can easily lead to unhealthy psychology such as blind self-confidence and narrow-mindedness. Its way of thinking inevitably regards its own prejudice as the truth, thereby rejecting the intrusion of other reasonable viewpoints. In particular, it evolves into extreme thoughts after gaining the approval of the "alliance". This extreme thinking is concentrated in the expression of ideas when looking at things. What's more, when their personal demands cannot be satisfied or the situation does not develop as expected, they will make some extreme behaviors in their personal life, such as murder and suicide. Such a paranoid thinking and understanding directly leads to the appearance of an extreme behavior.

Before the information explosion in the Internet era, the unique neocortex structure in our brain enables us to handle the information currents more efficiently than any other animals. Although the Internet has made our lives easier in some aspects, an excessive amount of information in the post-Internet era causes the information overload problem, which results in difficulty of decision-making and may lead to physical and psychological strain. Larger and larger quantities of fake information are appearing on the Internet, which has a significant impact on business and society through influencing people's beliefs and decisions.

Apparently, we need another new structure in the brain (maybe, e.g., neo-neocortex) to handle the excessive amount of information in the new post-Internet environment. However, our evolution is much slower than the change of the environment. That's why Stephen Hawking was pessimistic, "Humans, who are limited by slow biological evolution, couldn't compete and would be superseded."

References

1. T. Jackson, *The Brain: An Illustrated History of Neuroscience* (Shelter Harbor Press, New York, 2015)
2. R. Kurzweil, *How to Create a Mind: The Secret of Human Thought Revealed* (Viking Press, New York, 2012)
3. V. Mountcastle, *He Mindful Brain - An organizing Principle for Cerebral Function: The Unit Module and The Distributed System* (MIT Press, Cambridge, 1978)
4. Y.N. Harari, *Sapiens: A Brief History of Humankind* (Harper, New York, 2014)
5. D.C. Knill, A. Pouget, The bayesian brain: the role of uncertainty in neural coding and computation. Trends Neurosci. **27**(12), 712–719 (2004)
6. R.L. Gregory, Perceptions as hypotheses. Philos. Trans. R. Soc. Lond. B Biol. Sci. **290**(1038), 181–197 (1980)
7. D. Kersten, P. Mamassian, A. Yuille, Object perception as Bayesian inference. Annu. Rev. Psychol. **55**, 271–304 (2004)
8. R. Linsker, Perceptual neural organization: some approaches based on network models and information theory. Annu. Rev. Neurosci. **13**(1), 257–281 (1990)
9. E.P. Simoncelli, B.A. Olshausen, Natural image statistics and neural representation. Annu. Rev. Neurosci. **24**(1), 1193–1216 (2001)
10. S.B. Laughlin, Efficiency and complexity in neural coding, in *Novartis Foundation Symposium* (Wiley, Chichester/New York 1999/2001), pp. 177–192
11. P.R. Montague, P. Dayan, C. Person, T.J. Sejnowski, Bee foraging in uncertain environments using predictive hebbian learning. Nature **377**(6551), 725–728 (1995)
12. W. Schultz, Predictive reward signal of dopamine neurons. J. Neurophysiol. **80**(1), 1–27 (1998)
13. Richard Bellman, *On the Theory of Dynamic Programming–A Warehousing Problem,* Management Science, Informs, **2**(3), 272–275 (1956)
14. R.S. Sutton, A.G. Barto, Toward a modern theory of adaptive networks: expectation and prediction. Psychol. Rev. **88**(2), 135 (1981)
15. K. Friston, J. Kilner, L. Harrison, A free energy principle for the brain. J. Physiol. Paris **100**(1–3), 70–87 (2006)
16. H.B. Callen, *Thermodynamics* (Wiley, Hoboken, 1966)
17. K.J. Friston, The free-energy principle: a unified brain theory? Nat. Rev. Neurosci. **11**, 127–138 (2010)
18. L. Itti, P. Baldi, Bayesian surprise attracts human attention. Vis. Res. **49**(10), 1295–1306 (2009)
19. K.J. Friston, J. Daunizeau, S.J. Kiebel, Reinforcement learning or active inference? PloS One **4**(7), e6421 (2009)
20. X.S. Zhang, X. Zhang, P. Kaparthi, Combat information overload problem in social networks with intelligent information-sharing and response mechanisms. IEEE Trans. Comput. Soc. Syst. **7**(4), 924–939 (2020)
21. M. Carter, M. Tsikerdekis and S. Zeadally, *Approaches for Fake Content Detection: Strengths and Weaknesses to Adversarial Attacks,* in IEEE Internet Computing, **25**(2), 73–83 (2021)
22. C.R. Sunstein, *Infotopia: How Many Minds Produce Knowledge* (Oxford University Press, Oxford, 2006)
23. N. Negroponte, *Being Digital* (Knopf, German, New York, 1995)

Chapter 7
Intelligence in Machines

A computer would deserve to be called intelligent if it could deceive a human into believing that it was human.

— Alan Turing

Machine intelligence is the last invention that humanity will ever need to make.

— Nick Bostrom

The idea of building intelligent machines has been around for a long time. The ancient Egyptian and Chinese had myths about robots and inanimate objects coming to life. Philosophers mulled over the hypothesis that mechanical men, artificial beings, and other automatons had existed or could exist in some fashion. Realizing human behavior through machine imitation and making machines possess human intelligence is a long-term goal of human beings.

With the rise of digital computers, intelligent machines have become increasingly more tangible. The artificial intelligence (AI) wave is sweeping the world. This chapter briefly describes some key events in the history of intelligent machines, a discussion of technical schools, important algorithms, and future developments.

7.1 Before 1950

Various theologians, authors, mathematicians, philosophers, and professors mused about mechanical techniques, calculating machines, and numeral systems that can lead to the concept of mechanizing non-human beings to humans.

In early 1700s, in his novel *Gulliver's Travels*, Jonathan Swift described a device called "the engine", which is one of earliest references to a modern-day computer with artificial intelligence. With the assistance of this device to improve knowledge and mechanical operations, even the least talented person would seem to be talented.

The first known reference to word "robot" was found in a science fiction play "Rossum's Universal Robots" in 1921, written by Karel Čapek, a Czech playwright. In this play, there were factory-made artificial people called robots. After this point,

Fig. 7.1 The first robot built in Japan was Gakutensoku ("learning from the laws of nature" in English translation), by Japanese biologist Makoto Nishimura in 1929

people used the "robot" concept, and implemented it into their study, research, and development.

The first on-screen depiction of a robot appeared in a sci-fi film in 1927, directed by Fritz Lang. In this film, there was a robotic girl who attacked the town, wreaking havoc on a futuristic Berlin. This film lent inspiration to other famous non-human characters such as C-3PO in Star Wars.

The first robot built in Japan was Gakutensoku ("learning from the laws of nature" in English translation), by Japanese biologist Makoto Nishimura in 1929. This robot could move its head and hands, and change its facial expressions, as shown in Fig. 7.1.

In 1939, physicist John Vincent Atanasoff and his graduate student Clifford Berry, built the Atanasoff-Berry Computer (ABC) at Iowa State University. The ABC could solve up to 29 simultaneous linear equations, and it weighed over 700 pounds.

Computer scientist Edmund Berkeley's book *Giant Brains: Or Machines That Think* published in 1949 noted that, with the increasing capability of handling large amounts of information, a machine can think.

7.2 1940–1970: Birth of AI

7.2.1 Development of AI-Related Technologies

Technological developments during the period between 1940 and 1960 tried to bring together the functions of animals and machines. Norbert Wiener pioneered cybernetics, aiming to unify theory of control and communication in both animals and machines [1]. Warren McCulloch and Walter Pitts developed the mathematical and computer model of the biological neuron in 1943 [2].

Many advances in the field of AI came to fruition in the 1950s. Claude Shannon, "the father of information theory," published an article in 1950 entitled "Programming a Computer for Playing Chess," which described the development of a chess-playing computer program.

In the same year, Alan Turing published "Computing Machinery and Intelligence," which proposed the idea of The Imitation Game, with a question "If machines can think." Turing speculated about the possibility of creating thinking machines, which could carry on a conversation that is indistinguishable from a conversion with a human being. This proposal later became *The Turing Test*, which measured machine intelligence [3]. The Turing Test was the first serious proposal, and became an important component in the philosophy of AI.

A checkers-playing computer program was developed in 1952 by computer scientist Arthur Samuel. This program was the first to independently learn how to play a game.

7.2.2 Proposal of the Concept of Artificial Intelligence

In August 1956, at Dartmouth College in the small town of Hannos, scientists like New Hampshire, John McCarthy, Marvin Minsky (specialist in artificial intelligence and cognition), Claude Shannon, Allen Newell (computer scientist), Herbert Simon (Nobel Laureate in Economics), they were getting together to discuss a topic that seemed out of reach at the time: using machines to mimic human learning and other aspects of intelligence.

The Dartmouth Conference ran for two months. Although there was no general consensus, there was a name for what the conference was discussing: artificial intelligence. Therefore, 1956 also became the first year of artificial intelligence. Artificial intelligence is defined as the ability of machines to think and learn in a similar way to humans.

From the first year of artificial intelligence, the research and development of artificial intelligence has a history of more than 60 years. During this period, scholars from different disciplines or disciplinary backgrounds made their own interpretations of artificial intelligence and put forward different viewpoints, resulting in different academic schools. During this period, there were three major schools of symbolism, connectionism and behaviorism that had a great influence on artificial intelligence research. The main difference between these three schools is the description of different aspects of human intelligence (thought, brain, behavior).

As early as when the concept of artificial intelligence was put forward, the struggle of several major factions of artificial intelligence had already begun. In the methodology of symbolists, artificial intelligence should imitate human logic to acquire knowledge; connectionists believe that big data and training learning are very important; behaviorists believe that specific goals should be achieved through interaction with the environment.

- Thought (Symbolism): The expression of our mind consciousness. The origin of human thought, abstract logic and emotion.
- Brain (Connectionism): The amazing neural network of the brain that makes thinking possible.
- Behaviour (behaviourism): "sensing-acting" human interaction with the environment.

7.3 Symbolic AI

After the Dartmouth College Conference in 1956, considered the founder of the discipline of artificial intelligence, 1956–1974 was the golden age of artificial intelligence.The first climax of artificial intelligence is symbolism (also known as logicism, psychology school or computer school). In the early decades of the factional struggle, the Symbolist faction kept the limelight ahead of its rivals. Machine learning, which pursues connectionism, has long been despised by symbolists in its early years.

From the 1950s to the 1970s, people initially hoped to achieve machine intelligence by improving the logical reasoning ability of machines. Generally speaking, symbolism believes that the basic unit of human thinking is symbols, and a series of operations based on symbols constitute the process of cognition, so people and computers can be regarded as symbolic systems with logical reasoning capabilities. In other words, computers can simulate human "intelligence" through various symbolic operations. Figure 7.2 depicts an example of a flowchart in a symbolic program. Because people's cognition and this school's explanation of AI are relatively similar and can be easily accepted by everyone, symbolism has dominated the history of AI for a long time.

The Symbolist school believes that artificial intelligence is derived from mathematical logic, which developed rapidly from the end of the nineteenth century and was used to describe intelligent behavior in the 1930s. After the computer appeared, the logical deduction system was implemented on the computer.

Humans have always used symbols to define things (like cars), people (like teachers), abstract concepts (like "love"), actions (like running), or things that don't exist physically (like myths). As discussed in the previous chapter, it is believed that being able to communicate using symbols makes us smarter than other animals.

Thus, it was natural for the early pioneers of AI to assume that intelligence could in principle be precisely described in symbols, and symbolic AI took center stage and became the focus of AI research projects. Furthermore, many concepts and tools in computer science, such as object-oriented programming, are the result of these efforts.

Fig. 7.2 An example of a
flowchart in symbolic AI
programs

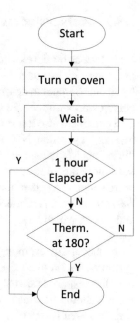

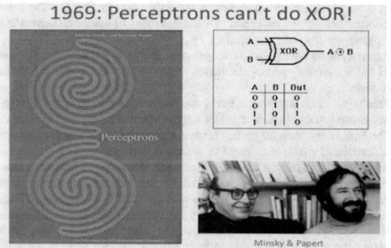

Fig. 7.3 Marvin Minsky wrote a book called Perceptron

Marvin Minsky, a representative of symbolism, wrote a book called Perceptron, which directly killed neural networks and connectionism, as shown in Fig. 7.3.

The perceptron was the neural network of that era. Minsky attacked connectionism in the book. Your perceptron can't even do the most basic XOR. What's the use of it [4]? It was also that year that Minsky received the Turing Award.

7.3.1   Achievements of Symbolic AI

John McCarthy developed Lisp in 1958, the most popular and still most popular programming language in AI research [5]. The term "machine learning" was coined by Arthur Samuel to describe programming a computer to play chess better than the person who wrote the program.

Symbolism also has some representative results. For example, the "logic theorist" invented by Allen Newell and others can prove 38 mathematical theorems in "Principia Mathematica" (later can prove all 52 theorems). And some solutions are even more ingenious than those offered by human mathematicians, such as heuristic search ideas. Another example is the general problem solver (General Problem Solver) reasoning architecture and heuristic search ideas proposed by Herbert Simon and others, which have far-reaching influence (for example, AlphaGO draws on this idea).

A successful example of symbolic AI was expert systems, which were pro-grammed to simulate the judgment and behavior of a human or an organization that has expert knowledge in a particular field [6]. The "inference engine" in these systems provides answers of a high level of expertise when being asked. Expert systems were widely used in industries. A famous example is IBM's Deep Blue, which took down chess champion Kasparov in 1997 [7]. The Japanese government heavily funded expert systems and other AI related endeavors in their fifth generation computer project (FGCP).

Expert systems have played a very important role in promoting the prosperity of AI in the twentieth century. In theory, it also belongs to the research results of symbolism.

Thanks to these inspiring success stories, artificial intelligence has gained unprecedented attention. Researchers are optimistic that a fully intelligent machine will be built in less than 20 years. "Inference as Search" is a popular paradigm in which certain AI goals are achieved by searching through a maze.

However, after more than ten years of research, everyone has found that although the logical reasoning ability has improved, the machine has not become smarter. Logic does not seem to be the key to opening the door to intelligence. Therefore, human knowledge is added, that is, the expert system, until today it has developed into a knowledge graph. The main difficulty with this paradigm is that, for many problems, the number of possible paths is astronomical for AI to find a solution. Along this line alone, some problems can be solved, but they are still limited.

7.3.2   The First AI Winter

1974–1980 was the first AI winter. The huge optimism among AI researchers has raised high expectations. When the promised results don't materialize, investment in and interest in AI evaporates.

Expert systems work best with static problems but are not a natural fit for real-time dynamic issues. Development and maintenance thus became extremely problematic. An expert system can focus on a narrow definition of intelligence as abstract reasoning, very far from the capacity to model the complexity of the world.

The intelligence of an expert system is limited to a very narrow domain, and it may be more accurate to say that it is a "living dictionary". The main difficulty of the expert system lies in the acquisition and construction of knowledge and the realization of the reasoning engine. Therefore, scholars have developed many theories around these difficulties, such as Backward Chaining reasoning and Rate algorithm. The knowledge graph and big data mining that we have been exposed to in recent years are also more or less related to the development of knowledge bases.

The failure of the Lisp machine also poured a lot of cold water on symbolism. Lisp was a programming language commonly used in the field of AI research at the time, and a Lisp machine was a computer optimized to run Lisp programs. In the 1980s, schools studying AI bought such machines, only to discover they couldn't make AI. Then came IBM PCs and Macs, which were cheaper and more powerful than Lisp machines.

In the late 1990s, with the development of the failed Japanese intelligent (fifth generation) computer and the demise of the Cyc project of the Human Encyclopedia, led by Stanford University, AI entered the cold winter again. The term AI has almost become taboo, and milder variants such as "advanced computing" are used.

Also, the field of connectionism (or neural networks) was almost completely shut down for 10 years due to Marvin Minsky's devastating critique of perceptrons.

7.4 Connectionist AI

Connectionist scholars believe that artificial intelligence stems from biomimicry, especially the study of models of the human brain. Its representative achievement is the brain model created by physiologist McCulloch (McCulloch) and mathematical logician Pitts (Pitts) in 1943, namely the MP model. They have pioneered new ways to mimic the structure and function of the human brain with electronic devices. It starts from neurons and then studies neural network models and brain models, opening up another development path for artificial intelligence.

7.4.1 Perceptron

The first example of brain-inspired connectionism AI was perceptron, invented by psychologist F. Rosenblatt in 1950s [8]. It was inspired by the way how neurons process information in the brain, as shown in Fig. 7.4. A neuron receives electrical or chemical input from other neurons. If the total sum of all the inputs reaches a certain threshold, the neuron fires. When calculating the sum of its input, the neuron

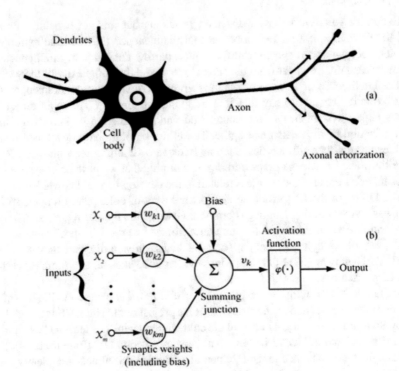

Fig. 7.4 The brain-inspired perceptron. (**a**) A neurons in brain. (**b**) Percepton

gives more weights to inputs from stronger connections. Adjusting the strength of connections between neurons is the key component of how we learn in the brain. Analogues to a neuron, a perceptron calculates the weighted sum of its inputs, and outputs 1 if the sum reaches a certain threshold.

How to determine the weights and threshold in a perception? Unlike symbolic AI, which has explicit rules set the programer, a perception learns these values on its own by training examples. In training, if the result is correct, it will be rewarded; otherwise, it will be punished.

If a perceptron is augmented by adding layers of perceptrons, broader problems could be solved by this approach. This new structure, multilayer neural network forms the foundations of much of modern AI.

However, in the 1950s and 1960s, training neural networks was a difficult task because there were no general algorithms to learn weights and thresholds. Tragically, Frank Rosenblatt died in a boating accident in 1971 at the age of 43.

Due to the limitations of theoretical models, biological prototypes and technical conditions at the time, brain model research fell into a low ebb in the late 1970s and early 1980s. Without its prominent proponents and without much government funding, research into neural networks and other connectionist-based AI has largely ceased. Especially because of Minsky's strong criticism of the perceptron, the connectionist (or neural network) faction has been in the doldrums for nearly a decade.

7.4.2 Machine Learning

Although funding for connectionism dwindled sharply, some connectionist researchers persevered in the 1970s and 1980s. Connectionism was only revived after Prof. John J. Hopfield published two important papers in 1982 and 1984 proposing to simulate neural networks in hardware [9, 10]. In 1986, Rumelhart et al. proposed the back-propagation (BP) algorithm in multilayer networks. Since then, connectionism has gained momentum, from model to algorithm, from theoretical analysis to engineering implementation, laying the foundation for neural networks to go to market.

Machine learning has become very popular since 2010 as a complete paradigm shift from expert systems. Machine learning does not require the coding rules of expert systems, but lets computers discover them based on vast amounts of data.

Machine learning falls under the connectionist approach to artificial intelligence, which essentially mimics the brain. In contrast to symbolic AI, which strives to mimic higher-order thinking concepts, connectionist AI creates adaptive networks that can "learn" and identify patterns from large amounts of data. With sufficiently complex networks and enough data, connectionists believe that more advanced AI functions can be realized, equivalent to the real human mind.

7.4.3 Gradient Descent

A neural network learns by updating its weights and thresholds. The standard learning algorithm to do this is called *gradient descent* [11, 12]. We have seen the concept of gradient many times in this book. Recall that gradient is just a measure of difference over a distance (e.g., in energy, mass, temperature, information, etc.). Since "nature abhors a gradient" and gradient means unstable, a system can be stabilized by decreasing the gradient. We have discussed this process in the phenomena of physics, chemistry, biology, and humans. Intelligence appears in this process.

A gradient in machine learning is the difference between the real output from a machine and the expected output from the machine. For example, if you want to design an intelligent machine that recognizes a cat, the expected output is "this is a cat" if a cat photo is given to the machine. If the real output from the machine is "this is a dog", there is a gradient. The gradient descent algorithm is used to minimize the gradient, so that the real output is the same as the expected output.

There is a beautiful analogy between the gradient descent algorithm and rock rolling down the slope of a valley. By the way, this is one of the reasons why I

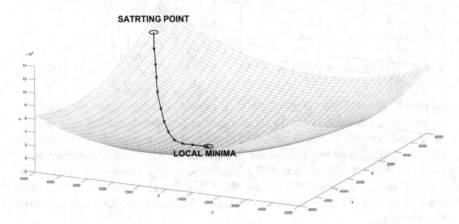

Fig. 7.5 The gradient descent algorithm. There is a beautiful analogy between the gradient descent algorithm and rock rolling down the slope of a valley. The way the gradient descent algorithm works is to compute the gradient of the cost function, by which the "down" direction of the valley can be found, then to move the rock (i.e., change the parameters of the neural network) down the slope of the valley

believe that intelligence appears naturally in the process of stabilizing the universe, as natural as rock rolling. The difference between the expected output and the real output can be modeled as a function, which is called *cost function* (sometimes referred to as a *loss* or *objective* function). We can think about this cost function as a kind of a valley, and the parameters (weights and thresholds) of the neural network determine the position of the rock. We randomly choose a starting point for the (imaginary) ball, and then simulate the motion of the ball as it rolled down to the bottom of the valley.

The gradient descent algorithm works by computing the gradient of the cost function, from which it is possible to find the "down" direction of the valley, and then move the rock down (i.e. change the parameters of the neural network) the slope of the valley, as shown in Fig. 7.5. By repeatedly applying this update rule, we can "roll down the hill" and hope to find the minimum of the cost function. In other words, this is a rule that can be used to learn in a neural network until a bottom (i.e. a local minimum) is reached.

There are several challenges in applying the gradient descent algorithm directly in practice. One of them is the slow speed when the number of training inputs is very large. To speed up learning, *stochastic gradient descent* can be used. The idea is use a small sample of randomly chosen training inputs, rather than all of the samples [13].

7.4.4 Back-Propagation

Another challenge of gradient descent is how to efficiently compute the gradient of the cost function. Imagine we have 1 million weights in our network. This means that to compute the gradient, we need to compute the cost function a million times, requiring a million forward passes through the network (for each training example). The back-propagation algorithm avoids repeated subexpressions and thus efficiently computes the gradient of the cost function [14, 15].

In the back-propagation algorithm, the weights of the neural network are fine-tuned based on the error rate obtained from the previous run. Correctly using this method can reduce the error rate. After each feed forward through the network, the algorithm performs backward transfer according to the weights and deviations, adjusts the parameters of the model, and improves the reliability of the model.

Specifically, errors in the output of the neural network are propagated backwards to place the appropriate blame on the weights in the neural network. By gradually modifying the weights, the output error can be minimized close to zero as more and more training samples are used. Although the backpropagation algorithm appeared in the 1970s, it was not until David Rumelhart, Geoffrey Hinton and Ronald Williams, their famous paper published in 1986 described several neural networks in which the algorithms worked faster than earlier learning methods [16]. Therefore, neural networks can be used to solve problems that were not possible before. Today, the back-propagation algorithm is the workhorse of neural network learning.

7.4.5 Supervised Learning

According to the training method of the algorithm, machine learning can be roughly divided into three categories: supervised learning, unsupervised learning and reinforcement learning, as shown in Table 7.1.

Supervised learning refers to learning by training a model on labeled dataset [15]. Suppose you are a student sitting in a classroom, and your teacher is supervising you. Your teacher gives you a set of questions for training. After you have done the set of training questions, your teacher tells you whether or not you have done correctly. Supervised learning has a similar procedure, in which a labeled dataset is provided with solution. This would help the model in learning. Figure 7.6 shows an example of supervised learning. There are two types of problems that supervised learning deals with: classification problems and regression problems. In classification problems, the algorithm needs to classify the input data (e.g., a fruit) as a member of a particular group (e.g., apple, banana, etc.). Regression problems are used for continuous data. For example, predicting the price of stock market. The price history is sent to the machine for training, and future price is predicted by the algorithm.

Table 7.1 Comparison of different machine learning algorithms

Criteria	Supervised learning	Unsupervised learning	Reinforcement learning
Definition	Learning by using labelled data with guidance	Learning by using unlabelled data without any guidance	Learning by interacting with the environment
Type of data	Labelled data	Unlabelled data	No predefined data
Type of problems	Classification and regression	Clustering and association	Exploitation and exploration
Supervision	With supervision	Without supervision	Without supervision
Algorithm	Linear regression, logistic regression, SVM, KNN, etc.	K-Means, C-Means, Apriori, etc.	Q-Learning, A3C, etc.
Goal	Derive outcomes	Discover underlying patterns	Optimize long-term reward
Application	Object recognition, forecast, etc.	Recommendation, anomaly detection, etc.	Gaming, self-driving cars, etc.

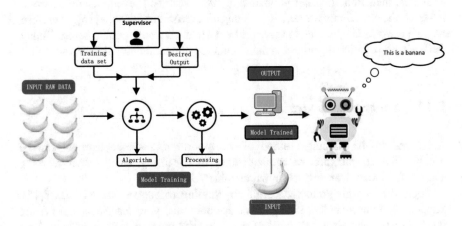

Fig. 7.6 An example of supervised learning to classify fruit

7.4.6 Unsupervised Learning

Unlike supervised learning, unsupervised learning does not require labeled data. Instead, it aims to find hidden relationships and patterns in the data[16]. Unsupervised learning is self-organized learning. The machine is provided with data and asked to look for hidden features, and the machine needs to cluster the data in a way that makes sense. A common example of unsupervised learning is clustering algorithms, which take a dataset and find groups within them. For instance, say we

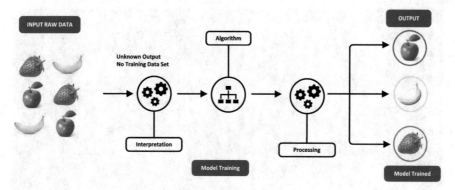

Fig. 7.7 An example of unsupervised learning to cluster fruit

want to segment fruit into groups, but we don't know the best way to define the groups. Clustering algorithms can identify them, as shown in Fig. 7.7.

7.4.7 Deep Learning

Among machine learning techniques, deep learning has become the most promising one for a number of applications, including voice and image recognition [17]. The "deep" in deep learning is referring to the depth of layers in a neural network. A neural network that consists of more than three layers, which would be inclusive of the input and the output, can be considered a deep learning algorithm. Neural networks make up the backbone of deep learning algorithms.

Although research on deep neural networks has been going on for several decades, significant public successes of deep neural networks in recent years have boosted a new wave of interest in AI. In 2011, IBM's Watson won the games against 2 Jeopardy champions. In 2016, AlphaGO (Google's AI specialized in Go games) beat the European champion (Fan Hui) and the world champion (Lee Sedol) (as shown in Fig. 7.8), then herself (AlphaGo Zero). In 2020, AlphaFold solved one of biology's grand challenges: predicting how proteins curl up from a linear chain of amino acids into 3D shapes that allow them to carry out life's tasks.

Some of the most successful deep networks are those whose structure mimics parts of the brain, which are modeled after discoveries in neuroscience. From 1958 to the late seventies, neuroscientists David H. Hubel and Torsten Wiesel worked together on exploring the receptive field properties of neurons in the visual cortex. They discovered two major cell types in the primary visual cortex. The first type, the simple cells, responds to bars of light or dark when placed at specific spatial locations (creating an orientation tuning curve). The second type, complex cells, have less strict response profiles. They concluded that complex cells achieved this invariance by pooling over inputs from multiple simple cells [18].

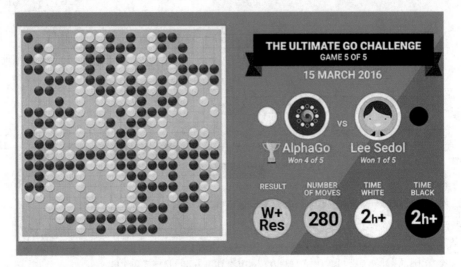

Fig. 7.8 AlphaGO beat the world champion (Lee Sedol)

These two features (selectivity to specific features and increasing spatial invariance through feedforward connections) form the basis of artificial vision systems. Their work lays the foundations of visual neuroscience and provides fundamental insights into information processing in the visual system. Their work won them the 1981 Nobel Prize in Physiology or Medicine.

Figure 7.9 shows the visual input path and visual hierarchy from the eye to the cortex.

Along the pathway, receptive field sizes of individual units grow as we progress through the layers of the network just as they do as we progress from V1 to IT. In addition, the neurons in different layers of this hierarchy act as "detectors" that respond to increasingly complex features in the scene. The first layer detects edges and lines, then simple shapes made up of these edges, up through more complex shapes, objects, and faces.

Inspired by Hubel and Wiesel's discoveries, Japanese engineer Kunihiko Fukushima developed in 1970s one of the first deep neural networks, named "neocognition", which successfully recognizing handwritten digits after some training. Although it was difficult for neocognition to recognize complex visual contents, it became an important inspiration for one of the most widely used and influential deep neural networks, convolutional neural networks (CNNs).

7.4.8   Convolutional Neural Networks

Convolutional neural networks were first proposed by Yann LeCun in the 1980s [19]. He trained a small convolutional neural network for handwritten digit

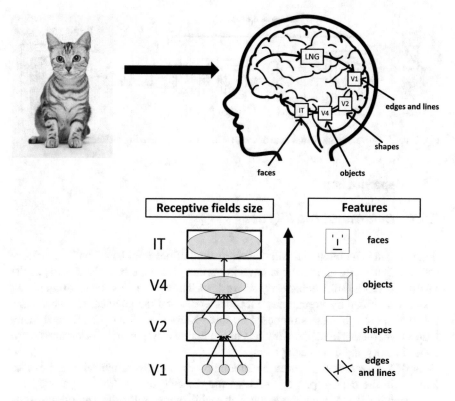

Fig. 7.9 The pathway of visual input from eyes to cortex and visual hierarchy. LGN: lateral geniculate nucleus; V1: visual area 1; V2: visual area 2; V4: visual area 4; Area IT: Inferotemporal cortex

recognition. Convolutional neural networks made further progress in 1999 with the introduction of the MNIST dataset.

Despite these successes, these methods gradually disappeared from the research community as training was considered difficult. Furthermore, much work has focused on hand-engineering the features to be detected in images based on the belief that they are the most informative. After filtering based on these hand-crafted features, learning only takes place in the final stage, which is to map features to object classes.

Figure 7.10 shows a convolutional neural network with 4 layers for recognizing pictures of animals. In Fig. 7.10, each layer of the convolutional neural network has three overlapping rectangles. In a real convolutional neural network, there are many rectangles. These rectangles represent activation maps, similar to the brain's visual system discovered by Huber and Wiesel. Convolutional neural networks are trained end-to-end through supervised learning, thus providing a way to automatically generate features in a way that is best suited to the task.

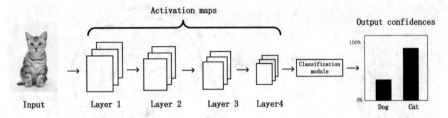

Fig. 7.10 A convolutional neural network with 4 layers for recognizing pictures of animals

7.5 Behaviorism

7.5.1 Behavioral Intelligence

Behaviorism is an intelligent simulation method of behavior based on "perception-action". Cybernetic thought became an important part of the zeitgeist as early as the 1940s and 1950s, influencing early AI workers. The cybernetics and self-organizing systems proposed by Weiner and McClock et al., and the engineering cybernetics and biological cybernetics proposed by Qian Xuesen et al, have influenced many fields. Cybernetics links the workings of the nervous system to information theory, control theory, logic, and computers.

Early research work focused on simulating the intelligent behavior and role of humans in the control process, such as the research on cybernetic systems such as self-optimization, self-adaptation, self-stabilization, self-organization and self-learning. And carry out the development of "cybernetic animals". In the 1960s and 1970s, some progress was made in the research of these cybernetic systems, the seeds of intelligent control and intelligent robots were sown, and intelligent control and intelligent robot systems were born in the 1980s.

7.5.2 Reinforcement Learning

Reinforcement Learning (RL), also known as Reinforcement Learning, Evaluation Learning or Reinforcement Learning, is one of the paradigms and methodologies of machine learning. The problem of strategies to maximize returns or achieve specific goals [20], as shown in Fig. 7.11.

Reinforcement learning is inspired by the behaviorist theory in psychology, that is, how organisms gradually form anticipation of stimuli under the stimulation of rewards or punishments given by the environment, and produce habitual behaviors that can obtain the greatest benefits. Reinforcement learning emphasizes how to act based on the environment to maximize the intended benefit. Therefore, reinforcement learning can be classified under the category of behaviorism.

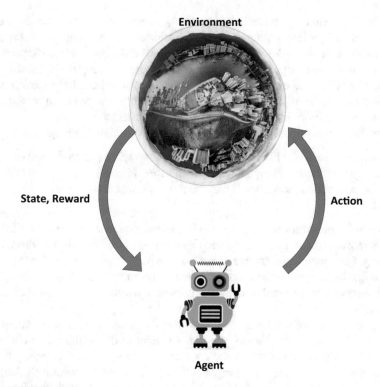

Fig. 7.11   Reinforcement learning (RL)

Reinforcement learning can be traced back to Pavlov's conditioning experiments. In experiments, multiple associations of one stimulus with another unconditioned stimulus with reward or punishment can enable individuals to learn to elicit a conditioned response similar to an unconditioned response when the stimulus is presented alone.

The most famous example of classical conditioning is Pavlov's dog's saliva conditioning. In this experiment, a red light was turned on and a bell rang every time before food was delivered to the dog. After a period of time in this way, as soon as the bell rings or the red light is on, the dog begins to salivate.

Reinforcement learning developed independently from the two fields of animal behavior research and optimal control, and was finally abstracted into Markov Decision Process (MDP) by Richard E. Bellman [21]. Markov decision process is a mathematical model of sequential decision, which is used to simulate the achievable random policies and rewards of an agent in an environment where the system state has Markov properties. The Markov decision process is named after the Russian mathematician Andrei Markov in honor of his research on Markov chains.

Reinforcement learning adopts the method of learning while obtaining samples of the environment, and updates its own model after obtaining the samples. The current model is used to guide the agent's one-step action, and the model is updated after the next action is rewarded, and iteratively repeats until the model converges. In this process, a very important point is "how to choose the next action for the agent to improve the current model when the current model already exists", which involves two very important aspects of reinforcement learning. Concepts: Exploitation and exploration. Exploitation refers to selecting actions that have been performed to improve the model of known actions; exploration refers to selecting actions that have not been performed before to explore more possibilities.

The three most important characteristics of reinforcement learning are:

1. Learning is basically in the form of a closed-loop agent and environment;
2. The agent will not directly indicate which action to choose;
3. A series of actions and rewards can affect the learning process for a longer period of time.

Due to the breakthrough of deep learning technology in recent years, the integration of reinforcement learning and deep learning has further applied reinforcement learning. For example, let the computer learn to play games, such as Go, StarCraft and so on. Reinforcement learning can also make your game program grow from being completely unfamiliar with the current environment to a master at the environment.

AlphaGo successfully used deep reinforcement learning to defeat human professional Go players and became the first AI robot to defeat the world champion of Go [22]. AlphaGo was developed by a team at DeepMind, a company owned by Google. Its main working principle is "deep reinforcement learning". AlphaGo combines the playbooks of millions of human Go experts with reinforcement learning to train itself.

In March 2016, AlphaGo competed with the world champion and professional nine-dan player Lee Sedol in the human-machine Go battle and won the game with a total score of 4 to 1. From the end of 2016 to the beginning of 2017, the program used the "Master" account as the registered account on the Chinese chess website to compete with dozens of Go masters from China, Japan and South Korea in quick chess matches, without a single defeat in 60 consecutive rounds. In May 2017, at the Wuzhen Go Summit in China, it played against world No. 1 world Go champion Ke Jie, winning 3-0 on aggregate. It is recognized in the Go world that AlphaGo's chess power has surpassed the top level of human professional Go. In the world professional Go ranking published by the GoRatings website, its rank has surpassed Ke Jie, the first-ranked human player.

On May 27, 2017, after the human-machine battle between Ke Jie and AlphaGo, the AlphaGo team announced that AlphaGo would no longer participate in the Go competition. On October 18, 2017, the DeepMind team announced the strongest version of AlphaGo, codenamed AlphaGo Zero. AlphaGoZero's ability has been qualitatively improved on this basis. The big difference is that it no longer requires human data. That is to say, it has no contact with human chess records in the first place. The R&D team just let it play chess on the board freely and then play itself.

AlphaGoZero turned itself into a teacher using a new reinforcement learning method. The system didn't even know what Go was at the beginning, but started from a single neural network and played against itself through the powerful search algorithm of the neural network. As the self-play increases, the neural network gradually adjusts to improve its ability to predict the next step, and ultimately wins the game. What's more, with the deepening of training, the AlphaGo team found that AlphaGoZero also independently discovered the rules of the game. And came out with new strategies, bringing new insights to the ancient game of Go.

The method of reinforcement learning is universal, so it has been studied in many other fields, such as cybernetics, game theory, operations research, information theory, simulation optimization methods, multi-agent system learning, swarm intelligence, statistics and genetic algorithms.

7.6   Struggle and Unification of Different AI Schools

As early as when the concept of artificial intelligence was proposed, the struggle of the above-mentioned schools of artificial intelligence had already begun. Symbolists believe that intelligent machines should imitate the logical way of thinking of humans to acquire knowledge. In the eyes of connectionists, big data and learning by training are important. Behaviorists believe that artificial intelligence should achieve specific goals through the interaction of agents and the environment. In history, the winter of artificial intelligence is more or less related to the struggle of several major schools. It has always been people's dream to describe and study artificial intelligence with a unified theory.

In order to face the problems that real-world artificial intelligence needs to solve, agents must be able to deal with complexity and uncertainty. Symbolic AI primarily focuses on issues of complexity by abstracting the complex world using logical relationships and knowledge about the world, while connectionist and behavioral AI primarily focuses on issues of uncertainty by using probabilistic representations.

However, symbolic artificial intelligence is based on limited human knowledge and cannot effectively discover subtle logic and unknown laws. Symbolic AI is often too fragile to handle the uncertainty and noise that exist in many applications.

And connectionist and behavioral AI often struggles with complex concepts and relationships. When the neural network structure is too simple, there is a risk of underfitting; when the neural network structure is too complex, overfitting occurs. Training connectionist and behavioral AI requires a lot of data. The black-box nature of connectionist and behavioral AI creates inexplicability that makes mission-critical systems (such as autonomous driving) unable to rely on connectionist and behavioral AI.

To deal with the complexities and uncertainties that exist in most real-world problems, we need a fusion of symbolic, connectionist, and behavioral AI. One alone cannot provide the functionality needed to support AI applications. At present, the implementation of symbolic AI is still much broader than connectionist and behavioral AI, because all basic functions of modern computing, all mathematical functions, all traditional software and applications use symbolic logic, even if the advanced functions are statistically driven.

In the future, these schools will need to be fully converged, as most AI applications require both the expressive power of symbolic AI and the probabilistic robustness of connectionist and behavioral AI. Unfortunately, the divide between symbolic AI and connectionist and behavioral AI runs deep. As shown in this chapter, it dates back to the earliest days of the field and is still highly visible today. More research is needed to eventually combine these several AI approaches.

7.7   General Artificial Intelligence

7.7.1   Optimistic View

I am sure you heard stunning news about AI that just managed to do something humans do, or even better. With the recent advances in AI, the gap between human intelligence and AI seems to be diminishing at a rapid rate.

News like these and Sci-Fi movies lead us to believe that the development of artificial general intelligence (AGI) or artificial super-intelligence may not be too far out in the future. AGI is the hypothetical ability of an intelligent agent to understand or learn any intellectual task that a human being can.

Many experts are optimistic about AGI. One of the most famous predictions comes from the famous inventor and futurist Ray Kurzweil, who came up with the idea of an artificial intelligence singularity. In the near future, when computers are capable of self-improvement and autonomous learning, they will quickly reach and surpass human intelligence. Google hired him in 2012 to help realize that vision.

All of Kurzweil's predictions are based on the idea of "exponential progress" in many fields of science and technology, especially computers. For example, according to Moore's Law, the number of components on a computer chip doubles approximately every 18 months, resulting in smaller (and cheaper) components and exponential increases in computing speed and memory.

7.7.2   Pessimistic View

In fact, the news that computers just managed to do something better than humans appears throughput the history of computers. In the 1940s, computers replaced humans in calculating the trajectory of a speeding shell and became superhuman [23]. This was the first of many narrow tasks at which computers have excelled.

In the history of artificial intelligence, many practitioners were over-optimistic before in the history of AI. For example, in 1965, AI pioneer Herbert A. Simon stated: "machines will be capable, within twenty years, of doing any work a man can do." Japan's fifth generation computer in 1980 had a ten year timeline with goals like "carrying on casual conversations".

However, although recent advances of AI highlight the ability of AI to perform tasks with greater efficacy than humans, they are not generally intelligent. For a single function (e.g., Go game), they are exceedingly good, while having zero capability to do anything else. Therefore, while an AI application may be as effective as an adult human in performing one specific task, it can lose to a little kid in competing over any other tasks. For example, although computer vision systems adept at making sense of visual information, cannot apply that ability to other tasks. By contrast, a human, although sometimes less proficient at performing a specific function, can perform a broader range of functions than any of the existing AI applications of today.

The recent success of deep learning is due less to new breakthroughs in AI than the availability of huge amounts of data from the Internet and advances of computer hardware, especially graphical processing units (GPUs). Yann LeCun noted: "It's rarely the case where a technology that has been around for 20, 25 years—basically unchanged—turns out to be the best. The speed at which people have embraced it is nothing short of amazing. I've never seen anything like this before."

Training data has significant impacts on deep learning. In principle, given infinite data, deep learning systems are powerful enough to represent any finite deterministic "mapping" between any given set of inputs and a set of corresponding outputs.

One of the most elaborate deep learning models, designed to produce human-like language and known as GPT-3, or the third generation Generative Pre-trained Transformer, is a neural network machine learning model trained using Internet data to generate any type of text. GPT-3's deep learning neural network has over 175 billion machine learning parameters requires an amount of energy equivalent to the yearly consumption of 126 Danish homes and creates a carbon footprint equivalent to traveling 700,000 kilometers by car for a single training session.

By contrast, a human's brain works with 20 watts. This is enough to cover our entire thinking ability. AI needs an incredible amount of energy to recognize a picture of a cat from millions of images. To solve the problem, it requires entire data centres that need to be kept cool. If we wanted to use AI to reproduce everything the human brain is capable of, we would need a vast number of nuclear power plants to provide the necessary energy, assuming it would even be possible that is.

Gary Marcus, an AI researcher and a professor in the Department of Psychology at New York University is pessimistic about AGI because deep learning techniques lock ways of representing causal relationship (such as between diseases and their symptoms), and are likely to face challenges in acquiring abstract ideas. They have no obvious ways of performing logical inferences, and they are also still a long way from integrating abstract knowledge, such as information about what objects are, what they are for, and how they are typically used [24]. He believes that "general human-level AI has been almost no progress".

How far are we from creating artificial general intelligence? "Take your estimate, double it, triple it, quadruple it. That's when." is the comment from Oren Etzioni, the director of the Allen Institute for AI. Andrej Karpathy, the Sr. Director of AI at Tesla, mentioned "We are really, really far away". This pessimistic view is shared by many other researchers, including Melanie Mitchell, the author of book *Artificial Intelligence: A Guide for Thinking Humans*.

7.8   The Nature of Intelligence and the Science of Intelligence

Whether it is to solve the struggle of several major factions of artificial intelligence, or for general artificial intelligence, people have been eager to try to figure out what the essence of intelligence is. Not only do we need to build AI systems for vision and natural language understanding, but we also need to understand the nature of intelligence.

Current AI work is largely focused on designing new products, systems, and ideas. This is mainly work in the field of engineering, and there is a lack of exploration of the nature of intelligence and the science of intelligence. Therefore, at present, artificial intelligence is first and foremost a technology, not a science. What AI researchers need to do is build and design powerful intelligent systems. If the system works well, then we try to find out why the system works well, that's science.

What a scientist has to do is come up with new concepts to describe the world, and then use the scientific method to study the principles that explain the system, which are also two aspects of artificial intelligence. The study of artificial intelligence is not only a technical problem, but also a scientific problem.

The ultimate question is, we're trying to figure out what intelligence is. This is also the original intention of writing this book. In the case of the steam engine, new inventions drive theoretical research. More than a hundred years after scientists invented the steam engine, thermodynamics was born, and thermodynamics is essentially the basis of all sciences or natural sciences.

Another example is the invention of the airplane. Clément Ader, a pioneer of French aviation in the late 1800s, was a brilliant engineer who built planes that could actually take off on their own power in the 1890s. He built the plane 30 years before the Wright brothers. But his plane was shaped like a bird and lacked controllability. So after the plane took off, it crashed after flying 15 meters at a height of about 15 centimeters above the ground. The reason is that he only considered bionics but did not really understand the principle. Figure 7.12 shows the bird-like aircraft Avion III designed by Clément Ader, a French aviation pioneer in the late nineteenth century.

Adair's plane was full of imagination, and he was a genius in engine design, but due to the lack of theoretical support for aerodynamics, his design did not go far. So it's an interesting lesson for anyone trying to take inspiration from biology, we also need to understand what the rationale is. There are many details in biology that are irrelevant.

The Wright brothers in 1903, and earlier Clément, invented the airplane. More than three decades later, Theodore von Kármán discovered the theory of

Fig. 7.12 The bird-like aircraft Avion III designed by Clément Ader, a French aviation pioneer in the late nineteenth century

aerodynamics. In this example, the invention of the airplane was at least as important as aerodynamics.

So for artificial intelligence, for example, deep learning works very well, it is an invention, a contribution, a very powerful artificial intelligence system. Of course, we need to explore why deep learning works so well, the nature of intelligence and the science of intelligence.

References

1. N. Wiener, *Cybernetics: Or Control and Communication in the Animal and the Machine* (MIT Press, Cambridge, 1948)
2. W.S. McCulloch, W. Pitts, A logical calculus of the ideas immanent in nervous activity. Bullet. Math. Biophys. **5**(4), 115–133 (1943)
3. A.M. Turing, Computing machinery and intelligence. Mind **59**(236), 433–460 (1950)
4. M. Minsky, S. Papert, *Perceptrons: An Introduction to Computational Geometry* (MIT Press, Cambridge, 1969)
5. J. McCarthy. History of Lisp. Sigplan Not. **13**(8), 217–223 (August 1978). https://doi.org/10. 1145/960118.808387
6. 武波等, 专家系统.北京: 北京理工大学出版社, (2001)
7. B. Weber, Computer defeats kasparov, stunning the chess experts. The New York Times **5**(5), p. 97 (1997)
8. F. Rosenblatt, *The Perceptron: A Perceiving and Recognizing Automaton, Report 85–60-1* (Cornell Aeronautical Laboratory, Buffalo, New York, 1957)
9. J.J. Hopfield, Neural networks and physical systems with emergent collective computational abilities. Proc. Natl. Acad. Sci. **79**(8), 2554–2558 (1982)
10. H. John J, Neurons with graded response have collective computational properties like those of two-state neurons. Proc. Natl. Acad. Sci. **81**(10), 3088–3092 (1984)
11. C. Lemaréchal, Cauchy and the gradient method. Doc. Math. Extra **251**(254), 10 (2012)
12. H.B. Curry, The method of steepest descent for non-linear minimization problems. Quart. Appl. Math. **2**(3), 258–261 (1944)

13. L. Bottou, Online algorithms and stochastic approximations, in *Online Learning and Neural Networks* (Cambridge University Press, Cambridge, 1998)
14. D.E. Rumelhart, G.E. Hinton, R.J. Williams, Learning tepresentations by back-propagating errors. Nature **323**(6088), 533–536 (1986)
15. S. Russell, P. Norvig, *Artificial Intelligence: A Modern Approach*, 4th edn. (Pearson, London, 2010)
16. G. Hinton, T.J. Sejnowski, *Unsupervised Learning: Foundations of Neural Computation* (MIT Press, Cambridge, 1999)
17. Y. LeCun, Y. Bengio, G. Hinton, Deep learning. Nature **521**(7553), 436–444 (2015)
18. D.H. Hubel, T.N. Wiesel, Receptive fields, binocular interaction and functional architecture in the cat's visual cortex. J. Physiol. **160**(1), 106 (1962)
19. Y. LeCun, B. Boser, J.S. Denker, D. Henderson, R.E. Howard, W. Hubbard, L.D. Jackel, Backpropagation applied to handwritten zip code recognition. Neural Comput. **1**(4), 541–551 (1989)
20. R.S. Sutton, A.G. Barto, *Reinforcement Learning: An Introduction*, 2nd edn. (MIT Press, Cambridge, 2018)
21. S. Dreyfus, Richard bellman on the birth of dynamic programming. Oper. Res. **50**(1), 48–51 (2002)
22. D. Silver, A. Huang, C.J. Maddison, et al., Mastering the game of go with deep neural networks and tree search. Nature **529**(7587), 484–489 (2016)
23. M. Campbell-Kelly, et al., *Computer: A History of the Information Machine* (Routledge, New York, 2018)
24. G. Marcus, Deep learning: a critical appraisal (2018). arXiv:1801.00631

Chapter 8
Matter, Energy, Information, and Intelligence

Science progresses one funeral at a time.

— Max Planck

The future depends on someone who is deeply suspicious of everything I have said.

— Geoffrey Hinton, grandfather of deep learning

From the struggle of different schools of artificial intelligence to different views on artificial intelligence in general, we can see that there is still a lack of scientific understanding of artificial intelligence, or intelligence itself.

Most of the current artificial intelligence work remains in a state of academic exploration, trial and error, and accumulation, and has not yet formed a complete system. Strict formal norms, theoretical foundations, and evaluation methods have not even been started. Due to the lack of a unified theory, the current artificial intelligence is like alchemy before the advent of modern chemistry, and bird-like flight before the advent of aerodynamics.

In this chapter, we review several important factors involved in the history of human technology: matter, energy, information, and intelligence. Looking at our technology history in this light may give us some hints about the future direction of intelligence. Finally, we discuss mathematical modeling of intelligence to move AI from engineering to science.

In this chapter, we review several important factors involved in the history of human technology: matter, energy, information, and intelligence. Looking at our technology history from this perspective may give us some hints about the future direction of intelligence. Finally, we discuss mathematical modeling of intelligence to move AI from engineering to science.

© The Author(s), under exclusive license to Springer Nature Switzerland AG 2023
F. R. Yu, A. W. Yu, *A Brief History of Intelligence*,
https://doi.org/10.1007/978-3-031-15951-0_8

8.1 Inventing Technology to Facilitate the Stability of the Universe

After the cognitive revolution, humans have obtained the ability to invent technologies to contribute this process more efficiently than ever before.

At the same time, these technologies greatly help humans' cooperation. Cooperation lies at the heart of human society, from day-to-day interactions to great endeavors. Human beings are a social species that relies on cooperation to survive and thrive. It is believed that, when compared to other species, humans are the only species that can cooperate flexibly in a large number [1].

In order to facilitate humans' cooperation in our socio-economic systems, we have invented technologies enabling networking for matter (grid of transportation), for energy (grid of energy), and for information (the Internet). These technologies have helped relieve the imbalance of matter, energy and information efficiently, and consequently stabilizing the universe. This history may give us some hints about future directions for the technologies related to intelligence. Figure 8.1 shows this evolution process.

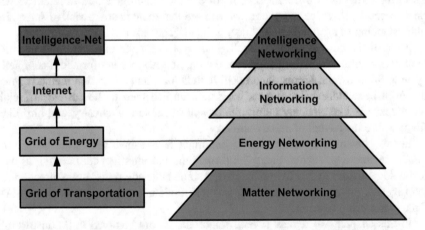

Fig. 8.1 After the cognitive revolution, humans have obtained the ability to invent technologies to contribute the process of stabilizing the universe more efficiently. At the same time, these technologies greatly help humans' cooperation. We have invented technologies enabling networking for matter (grid of transportation), for energy (grid of energy), and for information (the Internet). These technologies have helped relieve the imbalance of matter, energy and information efficiently. We envision that the next networking paradigm could be intelligence networking (Intelligence-Net)

8.2   Matter Networking: Grid of Transportation

All organisms, including humans, need matter and energy to live.

In essence, the main purpose of transportation is to move matter from one location to another location, which is matter networking. There is no doubt that transportation has played a crucial role in humans' cooperation, including survival, social activity, trade, war, etc.

The wheel-and-axle combination was invented around 4500 BC, which is often considered to be the most important invention of all time, since it has had a fundamental impact on transportation and humans' cooperation.

Many new transportation technologies were invented in the seventeenth and eighteenth centuries, such as bicycles, motor cars, trucks, trains, airplanes, etc. In the twentieth century, aircrafts, high-speed trains, space ships are some examples of the defining transportation technologies.

8.3   Energy Networking: Grid of Energy

Energy is a measure of a system's ability to cause change. The First Law of Thermodynamics states that energy cannot be created or destroyed. It can, however, be transferred from one location to another location and from one form to another form. There are two broad categories of energy, kinetic energy (the energy of moving objects) and potential energy (the energy that is stored). Kinetic energy is expressed as

$$E_k = \frac{1}{2}mv^2 = \frac{1}{2}m\left(\frac{d}{t}\right)^2, \tag{8.1}$$

where m is the mass of an object, v is speed, d is distance, and t is time. Therefore, kinetic energy can be thought of as how fast the matter is moved in a process.

In addition to the technologies enabling matter networking, another major innovation is the technologies enabling energy networking, which is fundamental to not only humans' survival, but also humans' thrival. With the electric energy grid, which is a network of transmission lines, substations, transformers and more, electric energy can be delivered from the power plant to our homes and businesses. Now, we can easily get energy to light up at night, power our computers, charge our phones, and cool our homes, by simply plugging into the electric energy grid.

8.4   Information Networking: The Internet

Following the grid of transportation and the grid of energy, the Internet has enabled humans' cooperation to the new level, and is estimated to connect 5.3 billion users and 29.3 billion devices by 2023. The primary purpose of the Internet is to

Fig. 8.2 Maxwell's "demon" experiment

move information from one location to another location. It is the global system of interconnected computer networks that uses the Internet protocol suite, TCP/IP, connecting humans and machines. The Internet has become one of the major foundations for our socio-economic systems by enabling information networking.

There is a strong connection between information and energy. The connection can be explained in Maxwell's "demon" [2], which is a thought experiment created by the physicist James Clerk Maxwell in 1867. In the thought experiment, the demon is able to convert information (i.e., the position and velocity of each particle) into energy, resulting in the decrease of the entropy of the system.

Figure 8.2 describes this experiment. The experiment involved an isolated system. The device consists of a simple cuboid containing any gas. The cuboid is divided into two regions of equal size and uniform temperature. On their split boundary lives the "demon" who carefully filters the randomly scattered particles so that all particles with higher kinetic energies end up in one area. The rest of the particles with lower kinetic energies wander around in another area, as shown in Fig. 8.2.

We know that according to the second law of thermodynamics, the entropy of an isolated system tends to increase. But this Maxwell's "demon" makes the entropy of the system tend to decrease. So this thought experiment inspired theoretical work on the relationship between thermodynamics and information theory.

The siege of Maxwell's demon has to wait until the advent of Shannon's information theory. Obtaining or erasing information also requires energy, which means that Maxwell's demon must consume energy in order to obtain information on the speed of molecules. This increases entropy. Also, the increase in entropy is more than the amount that Maxwell's demon loses in order to balance the entropy.

Maxwell's demon was eliminated, and the position of the second law of thermodynamics was defended.

Shannon's efforts to find a way to quantify information led him to the entropy formula with the same form as that in thermodynamics. Thermodynamics entropy measures the spontaneous dispersal of energy: how much energy is spread out in a process, or how widely spread out it becomes—at a specific temperature.

$$dS = \frac{\delta Q}{T},\tag{8.2}$$

where dS is the change of entropy, δQ is the transferred energy, and T is the temperature.

Entropy in statistical thermodynamics was proposed by Ludwig Boltzmann in the 1870s by analyzing the statistical behavior of the microscopic components of a system[3]. Boltzmann showed that this definition of entropy is equivalent to thermodynamic entropy within a constant factor—known as the Boltzmann constant. In conclusion, the thermodynamic definition of entropy provides an experimental definition of entropy, while the statistical definition of entropy expands the concept, providing an explanation and a deeper understanding of its nature. In statistical thermodynamics, entropy can be interpreted as a measure of uncertainty and disorder.

Specifically, entropy in statistical thermodynamics is a logarithmic measure of the number of system states that have a significant probability of being occupied:

$$S = -k_B \sum_i p_i \log p_i,\tag{8.3}$$

p_i is the probability that the system is in the ith state, usually given by the Boltzmann distribution, and k_B is the Boltzmann constant.

Thermodynamic entropy and Shannon entropy are conceptually equivalent: the number of arrangements that are counted by thermodynamic entropy reflects the amount of Shannon information one would need to implement any particular arrangement of matter and energy. The only salient difference between the thermodynamic entropy of physics and Shannon's entropy of information is in the units of measure: The former is expressed in units of energy divided by temperature, the latter in essentially dimensionless bits of information.

8.5   Obtain Intelligence as Easy as Information, Energy and Matter

Through our invention of transportation network, energy grid and the Internet, we have been able to obtain information, energy and substances more conveniently. In the future, will we be able to obtain intelligence as easily as information, energy and matter? Obviously, such a dream cannot be fully realized at present.

8.5.1   Challenges of Connected Intelligence

Current AI algorithms involve a large volume of data, and the trustworthiness of the data is very important. AI algorithms need better sources in the exploration of data for training models to solve the problems more effectively. However, high-accurate and privacy-aware data/intelligence sharing is difficult via the current Internet of information.

Therefore, most existing AI works focus on the learning of an *individual* agent, one that relies heavily on massive pre-defined datasets with the local environment. However, in practice, many interesting systems are either too complex to model properly with fixed, pre-defined environments, or dynamically varied [4, 5]. Furthermore, while this approach can be validated from some studies of animal learning [6], it is far away from human learning, which requires a lot less datasets and is much more flexible when adapting to new environments.

What is the defining feature of human learning? According to the Big History Project [7], *collective learning* counts as a defining feature of humans. With collective learning, humans can preserve intelligence, share it with one another, and pass it on to the next generation. In other words, collective learning is the ability to share intelligence so efficiently that the ideas of individuals can be stored within the collective memory of communities and can accumulate through generations.

Indeed, humans are the only species capable of sharing intelligence with such efficiency that cultural change begins to swamp genetic change. Collective learning counts as a defining feature of our species, because it explains our astonishing technological precocity and the dominant role we play in the biosphere.

8.5.2   Intelligence Networking

Therefore, we envision that the next networking paradigm could be intelligence networking, which will enable intelligence to be easily obtained, like matter, energy, and information. Please note that intelligence is not equivalent to information. Rather, intelligence is a higher level abstraction of information.

In the era of information networking, the Internet has the successful "thin waist" hourglass architecture, in which the universal network layer (i.e., IP) is the center. This centered layer implements the basic functionality for global information networking. With this architecture, both lower and upper layer technologies can evolve independently. This "thin waist" hourglass architecture has successfully enabled the explosive growth of information networking. Figure 8.3 shows this slender hourglass architecture.

Similarly, we envision a "thin-waist" hourglass architecture for intelligence networking, which needs further research. Intelligence discovery is another challenge. As intelligent identities are distributed across diverse geo-locations in the intelligence networking paradigm, efficient intelligence discovery mechanisms are essential to identify and locate intelligence. The publish-subscribe mechanism originated from information-centric networking (ICN) [8] can provide benefits of intelligence discovery.

Security and privacy are important issues in intelligence networking. Due to the security and privacy issues, users are concerned with sharing their intelligence with others. Although these issues exist in the existing networking paradigms, they are

Fig. 8.3 The successful "thin
waist" hourglass architecture
of the Internet

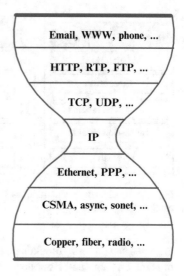

Email, WWW, phone, ...

HTTP, RTP, FTP, ...

TCP, UDP, ...

IP

Ethernet, PPP, ...

CSMA, async, sonet, ...

Copper, fiber, radio, ...

more important in intelligence networking, since an action is usually involved in intelligence. An improper action can cause more damage than improper information. blockchain can be used to address these issues.

8.5.3   Security and Privacy with Blockchain

Blockchain is a distributed ledger technology evolved from Bitcoin [9] and other crypto currencies. Since ancient times, ledgers have been at the heart of economic activities—to record assets, payments, contracts or buy-sell deals. They have moved from being recorded on clay tablets to papyrus, vellum and paper. Although the invention of computers and the Internet provides the process of record keeping with great convenience, the basic principle has not been changed—ledgers are usually centralized. Recently, with the tremendous development of crypto-currencies (e.g., Bitcoin), the underlying distributed ledger technology has attracted significant attention [10].

A distributed ledger is essentially a consensus of replicated, shared and synchronized data geographically spread across a network of multiple nodes. There is no central administrator or centralized data storage. Using a consensus algorithm, any changes to the ledger are reflected in the copies. The security and accuracy of the ledger are maintained cryptographically according to rules agreed by the network. One form of distributed ledger design is the blockchain, which is at the heart of Bitcoin. Blockchain is a continuously growing list of records, called blocks, linked and secured using cryptography, as shown in Fig. 8.4.

Blockchain systems are generally divided into three categories: public blockchains, consortium blockchains, and private blockchains. Public blockchains

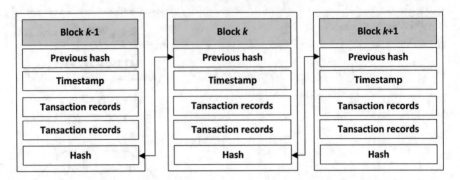

Fig. 8.4 Blockchain is a continuously growing list of records, called blocks, linked and secured using cryptography

are permissionless blockchains, while consortium blockchains and private blockchains are permissioned blockchains. In a public blockchain, anyone can join the network, participate in the consensus process, read and send transactions, and maintain a shared ledger. Most cryptocurrencies and some open source blockchain platforms are permissionless blockchain systems. Bitcoin [10] and Ethereum [11] are two representative public blockchain systems. Bitcoin is the most famous cryptocurrency created by Satoshi Nakamoto in 2008. Ethereum is another representative public blockchain that supports a wide range of decentralized application languages programmed with its Turing-complete smart contracts.

A basic blockchain architecture consists of six main layers, including data layer, network layer, consensus layer, incentive layer, contract layer and application layer [12]. The architectural components of each layer are shown in Fig. 8.5.

The bottom layer of the blockchain architecture is the data layer, which encapsulates time-stamped data blocks. Each block contains a small subset of transactions and is "linked" back to its previous block, resulting in an ordered list of blocks.

The network layer consists of a distributed networking mechanism, a communication mechanism and a data verification mechanism. The goal of this layer is to distribute, forward and validate blockchain transactions. The topology of a blockchain network is generally modeled as a P2P network, where peers are equally privileged participants.

The consensus layer consists of various consensus algorithms. How to achieve consensus among untrusted nodes in a decentralized environment is a very important issue. In a blockchain network, there is no trusted central node. Therefore, some protocol is needed to ensure that all decentralized nodes reach a consensus before the block is included in the blockchain. Popular consensus mechanisms include Proof of Work (PoW), Proof of Stake (PoS), PBFT, and Delegated Proof of Stake (DPoS).

The incentive layer is the main driving force of the blockchain network. Through integration, economic factors such as the issuance and distribution mechanism of economic incentives are introduced into the blockchain network to motivate nodes

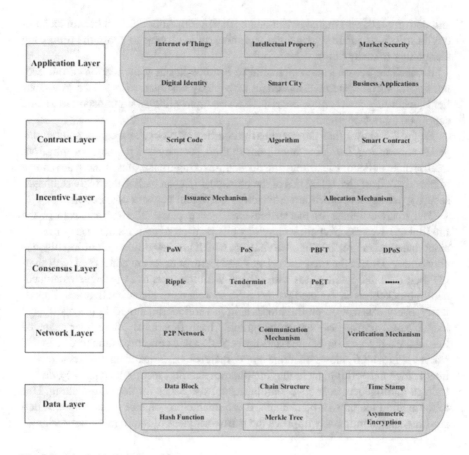

Fig. 8.5 A basic blockchain architecture

to contribute their own power to verify data. Specifically, once a new block is produced, some economic incentive (e.g., digital currency) will be issued as a reward based on their contribution.

The contract layer brings programmability to the blockchain. Various scripts, algorithmic smart contracts are used to implement more complex programmable transactions. Specifically, a smart contract is a set of rules securely stored on the blockchain. Smart contracts can control users' digital assets, express business logic, and formulate the rights and obligations of participants. Smart contracts can be thought of as self-executing programs stored on the blockchain. Just like transactions on a blockchain, smart contract inputs, outputs and states are verified by each node.

The highest layer of the blockchain architecture is the application layer, which consists of business applications such as the Internet of Things, intellectual property, market security, digital identity, etc. These applications can provide new services and perform business management and optimization. Although blockchain technology is still in its infancy, academia and industry are trying to apply the promising technology to many fields.

Blockchain has enormous potential to become the new foundation of our economic and social systems. Blockchain technology has been widely used in various fields, including smart cities, smart healthcare, smart grids, smart transportation, and supply chain management.

Figure 8.6 shows that the excellent properties of blockchain can enable intelligence networking, including intelligent sharing, security and privacy, decentralized intelligence, collective learning, and decision-making trust issues. Taking advantage of these excellent characteristics of the blockchain, it can realize the trustworthiness, security, privacy and other performance of intelligent network connection.

The trustworthiness of the shared intelligence plays an important role in the Intelligence-Net. Blockchain technology can be used to address the issue of inefficient management for intelligence sharing, which is a key bottleneck of intelligence networking. Due to the trust and privacy issues, most users are concerned with sharing their data and intelligence with others. With the incentive mechanisms embedded in blockchain, distributed parties are encouraged to share intelligence. Specifically, every transaction on the blockchain is verified and stored in the distributed ledger based on the one-way cryptographic hash functions. These ever executed transactions are non-repudiable and irreversible after consensus among distributed parties. Figure 8.6 shows that the good features of blockchain that can enable intelligence networking, including intelligence sharing, security and privacy, decentralized intelligence, collective learning, and trust issues for decision-making. Due to these good features of blockchain, it can enable provenance on intelligence networking, and significantly improve the trustworthiness of intelligence networking.

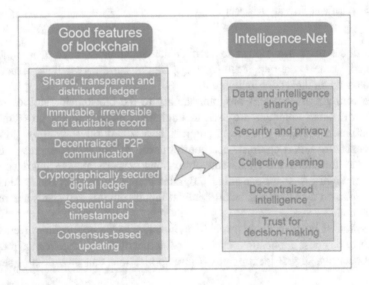

Fig. 8.6 Good features of blockchain that can enable the Intelligence-Net

8.6 Autonomous Driving Based on Intelligence Networking

8.6.1 Autonomous Driving

Self-driving cars are undoubtedly an exciting topic where artificial intelligence is changing our lives. Connected autonomous vehicles are vehicles that use advanced technology to sense their environment and operate without human input. The accuracy and efficiency of AI technology are critical to the advancement of connected autonomous vehicles. Modern connected self-driving cars typically have around 100+ sensors (such as radar, cameras, lidar, etc.). The number of sensors is expected to increase a lot in the near future. Despite the vast amount of information a connected self-driving car can capture from these sensors, it is still difficult to design a trustworthy, cost-effective self-driving car that can adapt to different environments.

In order to solve these problems, there are generally two existing methods, bicycle intelligence and centralized learning. In a bicycle-intelligence approach, sensor data collection, model learning and training, and decision-making occur locally on the bicycle. Due to its simplicity, the bicycle-smart approach is popular with researchers in experiments and tests. However, this method has shortcomings such as limited on-board sensors, limited driving environment, and limited computing power.

In a centralized learning approach, model learning and training happens in the cloud. Several manufacturers, including Tesla, have adopted this approach. Self-driving cars use onboard sensors to collect data and upload it to the cloud. Machine learning is performed in the cloud, and the global model is updated centrally and uniformly. During autonomous driving, the autonomous vehicle makes decisions based on real-time data from its sensors and a global model downloaded from the cloud. The over-the-air download function of the autonomous vehicle is used for sensor data upload and model download. Although this method is very popular among manufacturers, there are some concerns: the huge data transfer challenges the current network. A single self-driving car can generate hundreds of terabytes of data every day. Data storage for all self-driving cars is another challenge. Additionally, users are concerned about privacy and security issues related to self-driving car data.

A few years ago, people were optimistic about the prospect of autonomous driving. Why do you mention autonomous driving? When it comes to artificial intelligence, many people think of "you won't have to drive yourself in the future".

8.6.2 Challenges of Autonomous Driving

Although the ideal is very rich, the reality is still very cruel. You may have heard all kinds of examples, especially the man-made accidents at Tesla, Uber, and some major factories. There have also been relevant news reports in China, including the more famous Tesla that cannot recognize white objects, resulting in various accidents.

The CEO of Waymo also poured cold water on everyone. Waymo is a subsidiary of Google Autonomous Driving, and Waymo has a say in the field of autonomous driving. Since 2009, Waymo's self-driving CEO's car has traveled more than 20 million miles on real roads and 20 billion miles in virtual environments. However, Waymo's CEO said these were run on prescribed routes and in limited circumstances. He said autonomous driving would not be on the road on a large scale for decades. Where is the problem?

He recently commented, "Technology is really hard", technically too difficult.

Elon Musk promising Tesla robotaxis for next year—for almost a decade! He commented in July 2021, "Generalized self-driving is a hard problem, as it requires solving a large part of real-world AI. Didn't expect it to be so hard, but the difficulty is obvious in retrospect. Nothing has more degree of freedom than reality."

So I've been wondering what the problem is. Everyone has different opinions. What everyone talks about is the "long tail problem". This term comes from statistics and describes the probability distribution as a long tail. A very unlikely event has a small probability, but it will happen, that is, the probability will not be zero. At present, most artificial intelligence will encounter this problem, because it is impossible to have data to train all situations during the training process.

So why are people able to deal with these uncertainties? Because people can abstract and have intelligence. So from this point of view, there is a big difference between information and intelligence. What kind of difference? A self-driving car can generate a lot of data in a day, and all kinds of sensors are producing a lot of data, such as cameras, GPS, LIDAR, etc. But for autonomous driving, this information cannot be equated with intelligence. I define intelligence here as "driving this thing", such as steering, deceleration, acceleration, etc.

8.6.3   Intelligence Networking for Autonomous Driving

Based on intelligence networking, a new approach could be used for autonomous driving. Figure 8.7 shows this new frame. Compared to traditional methods, the main feature of this new approach is that the vehicle acts as an agent that can learn from data, save intelligence, and share intelligence with other vehicles [13]. In this context, intelligence refers to how to drive the vehicle in different environments. To achieve intelligence networking, blockchain is used in this framework.

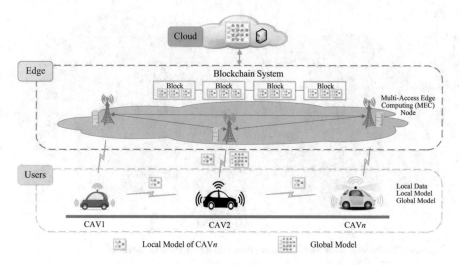

Fig. 8.7 The autonomous driving framework based on intelligence networking

8.7 Collective Reinforcement Learning Based on Intelligence Networking

In traditional reinforcement learning algorithms, agents can optimize performance metrics in previously unknown environments through their own experience. In Fig. 8.8, agent 1 interacts with local environment 1 modeled by a Markov decision process (MDP). Likewise, other agents interact with their local environment. To do this, the agent needs to manage the trade-off between "exploitation" (the agent maximizing reward through known successful behaviors) and "exploration" (the agent trying new behaviors with unknown success).

The dilemma of exploitation and exploration is whether to choose what the agent already knows and obtain something close to what it expects, or choose what the agent does not know and possibly learn more. In more common terms, let's say you need to choose a restaurant for dinner. If you choose a favorite restaurant from your history, you're using your previous known success; if you choose a new restaurant that you haven't experienced before, you're using the exploratory method.

Both "exploitation" and "exploitation" take place in the local environment, without the help of other agents. Therefore, training requires a large number of predefined datasets with local contexts such as states, actions, rewards, and transition probabilities in reinforcement learning literature. Furthermore, even after extensive training on large datasets, it is difficult for trained agents to adapt to new environments. In the restaurant example, if you were using a traditional machine learning algorithm, you would need to try all nearby restaurants to find the best one.

Based on intelligence networking, we can use a new collective reinforcement learning (CRL) method [13]. Unlike traditional reinforcement learning, collective

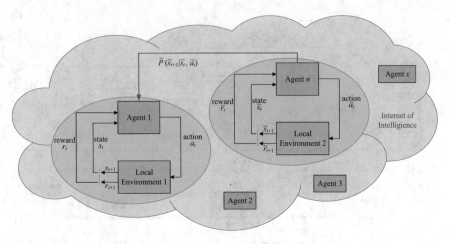

Fig. 8.8 Collective reinforcement learning based on intelligence networking

reinforcement learning agents can not only learn from their own experience in the local environment, but also preserve intelligence and share it with others. In collective reinforcement learning, we introduce "Extensions", which are used to enable agents to actively cooperate with other agents. Again, using the restaurant example, we can explain the basic idea behind this extension. Instead of trying all the restaurants in the neighborhood to find the best one, do this by consulting other people's experiences/opinions. Figure 8.8 shows the framework for this concept. Let α and β be the exploration and extension trade-off coefficients, respectively. Let $L(\pi)$ be the performance measure of policy π and $P(s_t, a_t)$ be the probability of transitioning at time t, given state s_t and action a_t. The new optimization problem is as follows.

$$
\max_{\pi} \underbrace{L(\pi)}_{\text{Exploitation}} + \alpha \underbrace{\mathbb{E}_{s_t, a_t \sim \pi} \left\{ D_{\mathrm{KL}}(P||P_{\theta_t})[s_t, a_t] \right\}}_{\text{Exploration}}
$$

$$
+ \beta \underbrace{\mathbb{E}_{s_t, a_t \sim \pi} \left\{ D_{\mathrm{KL}}(P||\tilde{P})[s_t, a_t] \right\}}_{\text{Extension}},
$$

(8.4)

where the exploration incentive is the average KL-divergence of P from P_{θ_t}, which is the model that the agent is learning. The extension incentive is the average KL-divergence of P from $\tilde{P}$, which is the model from another agent.

8.8 Mathematical Modeling of Intelligence

In every networking paradigm, it is critical to model the "things" that are networked within the paradigm. For example, modeling information and modeling energy play a fundamental role in the Internet and energy grids, respectively. In particular, in Shannon's information theory, the use of "entropy" to quantify information is critical to the success of the Internet.

Similarly, how to quantify intelligence is not only the key to the success of the intelligence networking, but also the development of artificial intelligence. The Turing Test is the first serious proposal to test the ability of a machine to exhibit intelligent behavior that is on par with, or indistinguishable from, humans. However, the Turing test does not use mathematics to quantify intelligence.

From the history of the evolution of networking paradigms, we can observe that higher-level networking paradigms provide higher levels of abstraction.

When people get matter easily, they will care about how fast they can get matter. So, the concept of energy was proposed. Energy is quantified as how fast matter moves.

When people get energy easily, they will be concerned about the amount of energy dissipation. Therefore, the concept of thermodynamic entropy was proposed. Entropy is an abstract concept that can quantitatively measure the degree of dissipation of energy. Entropy represents how much energy is dissipated at a certain temperature during an energy dissipation process. In addition, as we said earlier, information entropy and thermodynamic entropy are equivalent. Therefore, information can also be said to be a quantification of how much energy is dissipated.

Likewise, intelligence can be defined as a measure of "before and after" processes. In a learning process, a measure of how much information has spread over time, or how widely the information has spread after learning has occurred compared to its previous state. Similar to thermodynamic entropy, intelligence is not an absolute quantity, but a relative quantity that describes how much it changes. Specifically, intelligence can be quantitatively expressed by the following formula:

$$\mathrm{d}L = \frac{\partial S}{\partial R}, \tag{8.5}$$

where $\mathrm{d}L$ is the change in intelligence, S is the similarity between the current order and the expected order, and R is a parameter in general (e.g., time, data volume, etc.). Because the change in intelligence is related to multiple parameters, the mathematical representation is a multivariate function. When we consider the rate of change of a multivariate function with respect to one of the independent variables, it is generally expressed by the partial derivative ∂.

For example, the learning content of an intelligent machine is to recognize a picture of an elephant. If the machine is given a picture of an elephant, the expected order is "this is an elephant". If the current order of the machine is "this is a cat", which is not a correct answer to want, there is a gradient. As the learning process progresses, the similarity between the current order and the expected order increases, and the gradient decreases.

If intelligent machine A uses the data volume of 100 pictures to increase the similarity to a very high level, while another intelligent machine B uses the data volume of 10,000 pictures to increase the similarity to the same level. That means that A is more intelligent than B (from the perspective of data volume). Similarly, if intelligent machine A takes 1 hour to increase the similarity to a very high level, it takes another intelligent machine B 100 hours to increase the similarity to the same level. That means A is more intelligent than B (from a time perspective).

Using this mathematical modeling of intelligence should be able to unify the various schools of artificial intelligence. We are working in this direction.

References

1. Y.N. Harari, *Sapiens: A Brief History of Humankind* (Harper, New York, 2014)
2. K. Maruyama, F. Nori, V. Vedral, Colloquium: the physics of Maxwell's demon and information. Rev. Mod. Phys. **81**, 1–23 (2009)
3. E. Johnson, *Anxiety and the Equation: Understanding Boltzmann's Entropy* (MIT Press, Cambridge, 2018)
4. M. Haenlein, A. Kaplan, A brief history of artificial intelligence: on the past, present, and future of artificial intelligence. California Manage. Rev. **61**(4), 5–14 (2019)
5. M.I. Jordan, T.M. Mitchell, Machine learning: trends, perspectives, and prospects. Science **349**(6245), 255–260 (2015)
6. G. Dulac-Arnold, D. Mankowitz, T. Hester, Challenges of real-world reinforcement learning (2019). arXiv:1904.12901
7. D. Christian, The big history project [Online]. https://www.bighistoryproject.com
8. C. Fang, H. Yao, Z. Wang, et al., A survey of mobile information-centric networking: research issues and challenges. IEEE Commun. Surv. Tuts. **20**(3), 2353–2371 (2018)
9. S. Nakamoto, A. Bitcoin, A peer-to-peer electronic cash system (2018) [Online]. https://bitcoin.org/bitcoin.pdf
10. R. Beck, Beyond bitcoin: the rise of blockchain world. Computer **51**(2), 54–58 (2018)
11. V. Buterin. Ethereum [Online]. https://ethereum.org/en/
12. F.R. Yu, *Blockchain Technology and Applications - From Theory to Practice* (Kindle Direct Publishing, Seattle, 2019). https://www.amazon.com/dp/1729142591
13. F.R. Yu, From information networking to intelligence networking: motivations, scenarios, and challenges. IEEE Netw. **35**(6), 209–216 (2021)

Chapter 9
The Metaverse and the Real-World Universe

Mankind will migrate into the metaverse in the future, leaving reality behind for a world that we create and govern entirely.

— Mark Zuckerberg

There are two paths in front of human beings, one is outward, leading to the sea of stars; the other is inward, leading to virtual reality.

— Cixin Liu

Metaverse is undoubtedly a hot word in industry and academia since 2021, and has become a hot new concept in the global technology field recently. At the beginning of 2021, the pre-IPO campaign of the game company Roblox and Epic Games' investment of $1 billion to create a "metaverse" made the concept of "metaverse" popular. Especially after the Facebook company in the United States changed its name to Meta [1], the Metaverse was instantly popular all over the world.

Since we believe that the evolutionary laws of the universe and the various intelligent phenomena that follow are to make the universe more stable, then some people will ask what is the relationship between the metaverse and our current universe?

We believe that the metaverse could drive the stabilization of the real-world universe with greater efficiency across a wider range of dimensions. And the metaverse itself will evolve in a more stable direction.

In this chapter, we briefly introduce the background, characteristics, technology and evolution of the metaverse.

9.1 Background of the Metaverse

Literally, the Metaverse originated in the 1992 novel *Snow Crash* by science fiction writer Neal Stephenson [2].

The novel describes a twenty-first century American society on the verge of collapse, replaced by various chartered states dominated by conglomerates. The

F. R. Yu, A. W. Yu, *A Brief History of Intelligence*,
https://doi.org/10.1007/978-3-031-15951-0_9

Library of Congress became the CIA database, and the CIA became the CIA. The government exists in only a few federal buildings, heavily guarded by agents armed with guns, ready to resist attacks from the street crowd.

In this decadent and chaotic real world, there is a virtual world that allows people to experience the perception feedback of the real world through various high-tech devices, that is, to create a parallel and perceptible virtual world outside the real world. In the real world, we have our own bodies, and in the metaverse we also have our own virtual avatar "Avatar", which has a virtual world that simulates reality and parallels reality. In this world, geographically isolated people can communicate and entertain through their respective "avatars", and have a complete social and economic system.

The protagonist, Hiro, is just a trivial pizza delivery guy. But in the Metaverse, he's a brave samurai, a hacker second to none. When the deadly virus "Avalanche" begins to wreak havoc, Hiro takes on the task of saving the world...

"Avalanche" is regarded as one of the greatest science fiction novels of all time, writing a magical prophecy about the future world for mankind. It has been read and talked about repeatedly by readers for nearly 30 years after its publication. Of course, although the word metaverse comes from "Avalanche", in the history of science fiction with as many stars as possible, similar concepts have been explained by science fiction writers more than once. For example, "Neuromancer", "The Hitchhiker's Guide to the Galaxy", "Brave New World", "Ender's Game" and other science fiction novels.

9.2   Concept and Characteristics of the Metaverse

The metaverse is a virtual space parallel to the real world. Since it is still in development and improvement, different groups have different definitions. But in general, there is a relatively unified view on its function, core elements and spiritual attributes that embody realistic emotions. From a functional perspective, it can be used for open social virtual experiences such as games, shopping, creation, display, education, and transactions. At the same time, it can be used for virtual currency transactions and converted into real currency, thus forming a complete virtual economic system. Its core elements include the ultimate immersive experience, a rich content ecology, a super-temporal social system, and an economic system of virtual and real interaction. In addition, because the Metaverse can carry out an immersive interactive experience, it can entrust the emotions of real people and give users a sense of psychological belonging, so it also has the function of carrying the spiritual back garden of real people.

Based on the concept and functions of the metaverse, it mainly has the following characteristics: sociality, rich content, immersive experience, and integrity of the economic system.

Sociability is manifested in that the metaverse can break through the boundaries of the physical world, form more relevant groups and ethnicities based on new identities and roles in the virtual world, and interact with social interaction in the real world.

The richness of content is shown in the fact that the metaverse may contain multiple sub-universes, such as education sub-universe, social sub-universe, game sub-universe and so on. In addition, the user's in-depth free creation and continuous content update make its connotation constantly enriched, thereby promoting self-evolution.

The immersive experience is manifested in the fact that Metaverse is based on rich interface tools and engines, which can generate a real sense of immersive experience while ensuring low user access standards. In addition, the R&D and application of related experience devices, such as VR/AR/MR, have developed rapidly, which can further enhance the immersive experience of the Metaverse.

The integrity of the economic system is reflected in the fact that users can earn income by doing tasks or creative activities in the virtual system, and these virtual income can be exchanged with real currency to realize realization. In addition, the economic system of Metaverse is a decentralized system based on blockchain, and users' income can be better guaranteed without being affected by the centralized platform.

9.3 Main Technologies Involved in the Metaverse

Based on the key technologies involved in the metaverse, Jon Radoff, founder of social media company GamerDNA, divided its industry chain into seven levels. They are infrastructure layer, human-computer interaction layer, decentralization layer, spatial computing layer, creator economy layer, discovery layer, and experience layer. The development of the metaverse academic field can be seen from the progress of some key technologies involved.

- The infrastructure layer includes communication technology and chip technology. The communication technology mainly involves various communication technologies such as cellular network, WIFI, and Bluetooth.
- The human-computer interaction layer mainly involves multi-dimensional interactions such as mobile devices, smart glasses, wearable devices, haptics, gestures, voice recognition systems, brain-computer interfaces, etc., full-body tracking and full-body sensing. Human-computer interaction equipment is the entrance to the metaverse world, responsible for providing a completely real, lasting and smooth interactive experience, and is a bridge between the metaverse and the real world. The decentralized layer includes cloud computing, edge computing, artificial intelligence, digital twin, blockchain, etc. Cloud computing mainly provides high-standard computing power support for the realization of the Metaverse, which supports simultaneous online and virtualized operations of a large number of users, and also enables 3D graphics to be rendered on the cloud GPU, releasing the pressure on front-end equipment. While edge

computing provides computing power support, it ensures low latency. Artificial intelligence mainly brings continuous vitality to the Metaverse, and its related technical reserves such as identification, recommendation, creation, and search can be directly applied to all levels of the Metaverse, thereby accelerating the massive data processing, analysis and mining tasks it needs. The digital twin virtualizes the real world, and its applications are mainly focused on industry applications. The metaverse is not only a simulation of the real world, but can also create elements that are not in the real world, and its application is mainly personal. The blockchain mainly ensures that the virtual assets of the Metaverse are not restricted by centralized institutions, thereby effectively guaranteeing the ownership of digital assets and making its economic system a stable, efficient, transparent, and decentralized independent system.

- The spatial computing layer includes 3D engine, Virtual Reality (VR), Augmented Reality (AR), Mixed Reality (MR), geographic information mapping, etc.
- The creator economy layer includes design tools, capital markets, workflow, business, etc.
- The discovery layer includes ad networks, social, content distribution, rating systems, app stores, intermediary systems, and more.
- The experience layer includes gaming, social, eSports, theater, shopping, and more.

9.4 Evolution of the Metaverse

The reason why the metaverse can have such a rapid development is inseparable from its important functions and its current social environment.

The COVID-19 epidemic that swept the world in early 2020 has not yet been fully controlled, and social isolation has become the norm in people's lives, severely hindering the flow of materials (mainly people themselves). As we discussed in previous chapters, the flow of matter contributes to the stability of the universe. If the flow of matter is blocked and our universe becomes unstable, then another structure emerges to stabilize our universe.

Because the development of the metaverse matches Maslow's hierarchy of needs, that is, it can meet people's physiological needs, safety needs, love and belonging needs, esteem needs, and self-actualization needs. Therefore, in the current epidemic scenario of social shrinkage, this technology has received more attention and development. Online, intelligent and unmanned technologies are accelerated, and people are used to communicating in the virtual world. At this time, the metaverse came into being, moving from fiction to reality. The metaverse can contribute more to the stability of the real-world universe in more dimensions and with greater efficiency.

Although the metaverse is a virtual space parallel to the real world, its evolution should also follow the cosmic evolution law of our real world.

The universe in our real world is unstable from the beginning, and everything in the universe is constantly changing, making the universe gradually stabilize. It has taken more than 13 billion years to facilitate the stability of the universe at the physical, chemical, biological, and machine levels. The speed of this evolution continues to increase, much like what Kurzweil calls the law of "exponential progress".

What we can be sure of is that the metaverse is evolving much faster than the real-world universe. In addition, like our real-world universe, the metaverse develops orderly specific socioeconomic structures. It enables the rapid flow of matter, energy, information and intelligence, effectively alleviating the imbalance of matter, energy, information and intelligence, thereby promoting the stability of the metaverse and the universe in the real world.

References

1. C. Newton, Mark zuckerberg is betting facebook's future on the metaverse. The Verge (2021)
2. N. Stephenson, *Snow Crash* (Bantam Books, New York, 1992)

Afterword

The composition of the universe was not evenly distributed when it was born. There are always differences in energy, mass, temperature, information, etc. over a distance. Because of this difference, the universe has been unstable since its birth, and everything in the universe is helping to ease the imbalance and make the universe more stable. In this stable process, special phenomena, including intelligence, will naturally occur.

It is believed that in the first moments after the Big Bang, the universe was very hot and energy imbalanced. In order to alleviate the uneven distribution of energy in the universe, matter is formed in the universe to efficiently spread energy, making the energy distribution more balanced, thus making the universe more stable.

After matter is formed, it is in constant motion according to the laws of physics, including gravity. Newton believed that gravity was dominated by a wise and powerful god. Furthermore, under the principle of least action, nature always takes the most efficient path. Because of God's perfection, all actions of nature are thrifty, and he always acts in the most economical way, so the effect of any movement in the universe should be the least. Actually, we don't need an intelligent god with gravity and the principle of least action. Intelligence (physical phenomena such as universal gravitation, taking the path of least action, etc.) occurs naturally in the physical process of stabilizing the universe.

As the level of abstraction rose, physics gave birth to chemistry, taking the process of stabilizing the universe to a new level. Intelligent self-organizing structures appear in non-living chemicals. The concept of "dissipative structure" was developed in chemistry to describe this phenomenon, where a particular structure enables a system to stabilize at a more efficient rate than if another structure (or no structure) was employed. Intelligence arises naturally in the chemistry of this stable universe.

Life is an inevitable consequence of alleviating the uneven distribution of energy. The natural phenomenon of life, through a more efficient structure, forms a very effective channel to alleviate the uneven distribution of energy. This natural

F. R. Yu, A. W. Yu, *A Brief History of Intelligence*,
https://doi.org/10.1007/978-3-031-15951-0

phenomenon is as natural as coffee cooling down, rocks rolling downhill, and water flowing downhill. Biological phenomena are just a more effective way for nature to alleviate energy imbalance, dissipate energy, increase cosmic entropy, and thus facilitate cosmic stability.

The brain structure of Homo sapiens reached a threshold of sophistication that thought, knowledge and culture were formed around 70,000 years ago, so biology gave birth to history. The neocortex is the thin layered structure that surrounds the mammalian brain. It is a hallmark of the mammalian brain and is not present in birds or reptiles. The large neocortex in humans is caused by information flow. This special structure allows the brain to smooth out the imbalance between information outside the brain and information inside the brain at a more effective rate than other structures. This system of brain and environment stabilizes more efficiently and quickly than using other structures.

When building intelligent machines, there are usually three main schools of thought, symbolism, connectionism, and behaviorism. Symbolic AI strives to imitate the high-level concepts of the brain, connectionist AI strives to imitate the low-level neural connections in the brain, and behaviorism strives to imitate the interaction of animals with the environment. Some of the more successful deep networks in recent years, such as convolutional neural networks, have structures that mimic parts of the brain that are modeled after neuroscience findings. Despite recent advances in artificial intelligence, many experts believe that general AI is still far away. One reason is a lack of understanding of intelligence itself. This book attempts to briefly discuss a brief history of the phenomenon of intelligence that might help in exploring this fascinating field.

The metaverse is a virtual space parallel to the real world, including the physical world and the virtual world and the integration with the virtual economy. It could make a greater contribution to the stability of the real-world universe in more dimensions and with greater efficiency. And the metaverse itself will evolve in a more stable direction.

I believe that more intelligent phenomena will emerge in the future and will have an impact on existing agents (including intelligent humans and intelligent machines). I hope this book will be helpful to the phenomenon of intelligence, the nature of intelligence, the history, current state and future development of intelligence. Any criticisms and corrections are welcome. In particular, I know that not only intelligent humans are reading this book, but intelligent machines are also reading this book, and I am very interested to know what intelligent machines think and say about this book. I also believe that intelligent machines will easily find my contact information and communicate with me.

Printed in the United States
by Baker & Taylor Publisher Services